AF598260

Global Problems and Common Security

Global Problems and Common Security

Annals of Pugwash 1988

Edited by
J. Rotblat and V. I. Goldanskii

With 19 Figures

Springer-Verlag Berlin Heidelberg New York
London Paris Tokyo Hong Kong

Professor Dr. Josef Rotblat
Pugwash, Flat A, Museum Mansions,
63A Great Russell Street, London
WC1B 3BJ, United Kingdom

Academician Vitalii I. Goldanskii
Institute of Chemical Physics,
USSR Academy of Sciences,
Kosygina 4, 117334 Moscow, USSR

ISBN 3-540-51699-9 Springer-Verlag Berlin Heidelberg New York
ISBN 0-387-51699-9 Springer-Verlag New York Berlin Heidelberg

Printed in the United States of America

2157/3150-543210 – Printed on acid-free paper

Preface

After many years of sterile arms control negotiations between the super powers, which did not produce a single genuine disarmament measure – and sometimes had the opposite effect of stimulating the development of new weapons to serve as bargaining chips – an agreement to abolish completely two categories of nuclear weapons, the INF Treaty, was signed and is now being implemented. This historical event, the first actual destruction of deployed nuclear missiles (although not of their warheads) became possible largely because of the radical changes in the policies of the Soviet Union, the "new way of thinking", advanced by its leader Mikhail Gorbachev. The relaxation of tension that resulted from this policy was one of the factors contributing to the successful conclusion of the INF Treaty. Another, no less important factor, was the acceptance by the Soviet Union of on-site inspections as a basic element of a verification system essential to ensure compliance with the Treaty. This breakthrough led to the mutual acceptance of an elaborate and precise verification regime, and thus made possible the signing of the INF Treaty, which in turn contributed to the further lessening of tension between East and West.

But perhaps the most important factor, one which carries much promise for the future, was the changed approach to security problems. Until recently the policies of all nuclear-weapon states were based on the assumption that national security demands the possession of nuclear arsenals, larger, better, less vulnerable than those of the other side. The result of this was an unending arms race: steps taken by one side to improve its security were perceived by the other side as lessening *its* security, which in turn induced the latter to take other steps, seen by the first side as threatening its security, and so it went on and on. It was Mikhail Gorbachev who first broke out of this vicious circle by adopting the new way of thinking. In his book *Perestroika* he said: "there should be no striving for security of oneself at the expense of others." This epitomizes the radical change in approach, a change from the concept of national security to that of common security. This was the underlying philosophy of the Russell-Einstein Manifesto of 1955, which is the basis of the Pugwash Movement. President Gorbachev acknowledged the important roles of scientists when he said: "I think that it was a mistake when the people did not listen to the words of Einstein and his adherents. They have warned that the world has obtained a tremendous power which demands a completely new approach. At those times the reaction to this warning was, however, not sufficiently responsible, and now we are harvesting the results of respectless attitude towards the opinions of scientists." The recent statement by President George Bush on conventional arms is a hopeful sign that the new way of thinking is gaining wider support.

The implementation of the concept of common security is bound to bring further reductions in all types of armaments, and hopefully also treaties banning them altogether. The present enormous investment in arms – in terms of scientific manpower, technological resources, and capital expenditure – if thus released and applied to other areas which also constitute sources of conflict, could remove potential threats to the peaceful continuation of civilization. Notable among these areas are the degradation of the environment and the economic plight of the developing countries. There is a close link between these problems and security. We live in an interdependent world in which the major issues have to be tackled on a global scale.

The theme of the Pugwash Conference held in September 1988 in Dagomys, USSR "Global Problems and Common Security" – which is also the title of this volume – clearly stresses this link. The three major issues on the agenda of the Pugwash Movement, peace, pollution, and poverty, were the chief subjects of debate at the Dagomys Conference; nearly all the material for this book is papers commissioned for that Conference.

Because of the long-standing concentration of Pugwash on problems of security, and the considerable expertise acquired in this field, most of the book deals with these problems. Despite the relaxed political climate, security must still be given top priority. There is a great need to remind the public that the danger of a devastating war, particularly a nuclear war, is far from over. The INF Treaty is a great achievement, but it reduced the nuclear arsenals by only a few per cent, and this will be more than compensated for by the new weapons already deployed or being planned. There is still a long way to go before the threat of a military confrontation with nuclear, chemical, or conventional arms is effectively removed.

In *Part One* of the book the means to reduce the threat posed by the existing strategic nuclear arms are debated. In the first paper the need of drastic cuts in nuclear arsenals, much deeper than the 50 per cent now negotiated in START, is eloquently argued, and specific proposals are put forward for very deep reductions of those armaments. An important step towards the halting of the nuclear arms race is a comprehensive test ban: the verification regime needed for this purpose is outlined in the second chapter. The current attempts to achieve this by USA-USSR collaboration between scientists are described in the third paper. In the last chapter, the need is stressed to preserve the ABM Treaty in its original formulation, if a new spiral in the nuclear arms race is to be avoided.

Another danger arising from the existence of nuclear weapons, their spread to other states, is the subject of the papers in *Part Two*. The Non-Proliferation Treaty – with all its weaknesses – is still considered to be the best instrument to prevent the horizontal proliferation of nuclear weapons. The maintenance of the NPT is advocated by several authors, although one believes that the inadequacies of the Treaty call for its revision. There is a detailed discussion, in the last chapter, of an urgent proliferation issue – how to stop the spread of fissile materials, the products of the peaceful uses of nuclear energy, which might be converted to military purposes.

The reduction of nuclear arsenals is often linked with that of conventional armaments. The nature of that link is scrutinized in *Part Three*. The historical developments that have led to the policies of NATO and WTO are reviewed, and the prospects of the current negotiations on reductions in conventional arms are evaluated. In one paper the whole concept of nuclear deterrence is critically analysed, and ways are suggested to get away from reliance on it by a restructuring of conventional forces. The last chapter presents a summary of discussions on this topic in a special Pugwash Workshop held in Geneva in June 1988.

Part Four deals with the prevention of chemical warfare. Pugwash has a long history of concern with this issue, and its study has significantly contributed to the present climate favourable to an agreement on a Chemical Weapons Convention to prohibit not only the use but also the production and storage of such weapons. The first chapter, a review of the current situation, was a paper presented at a session of the Pugwash Workshop on Chemical Warfare held in Geneva in January 1988. The other two chapters outline the prospects for chemical disarmament, as seen by scientists in the UK and USSR.

Europe occupies a singular position in problems of security, being the area where NATO and WTO, both equipped with tactical nuclear weapons, face each other directly. Here is the greatest need for a restructuring of the military forces from an offensive to a defensive posture. *Part Five* contains detailed plans how to achieve this, how to introduce and maintain confidence-building measures, and how to verify them. The last chapter presents the views of the Pugwash Council on disengagement in Europe, in a statement issued after a Symposium on this topic held in Prague in April 1988.

Part Six deals with a problem considered by some to be at the root of the arms race, the role of military research and development. One paper, the author of which can speak from personal experience, stresses the vital part played by the technological innovators, while another paper takes a critical look at proposed ways of controlling military R&D. The economic aspects of conversion of military R&D to peaceful applications are discussed in the third chapter.

This leads directly to the link with problems of the environment and underdevelopment, the subjects of the last two parts of the book.

Part Seven tackles the global ecological problems and their relation to security problems. The first chapter examines the environmental consequences of a nuclear war and then goes on to an analysis of other ecological dimensions of security. The urgent need for new policies, if a catastrophic deterioration of the environment is to be avoided, is the conclusion arrived at in the second chapter, which contains an in-depth review of the climatic changes resulting from the injection of various chemicals into the atmosphere. This is developed further in the third paper which also stresses the potential for provoking disputes and international friction, unless mechanisms can be worked out to share more equitably the costs of climatic changes. Plans for international co-operation needed to ensure the survival of civilization are outlined in the last chapter.

A similar approach to the problems of underdevelopment is advocated in *Part Eight*, which is devoted to seeking ways to alleviate poverty, disease, and strife. The first chapter scrutinizes the close and multi-dimensional relationship between disarmament and development. The role of science and technology in alleviating underdevelopment is surveyed in the second paper. The special situation in Africa, the link between food shortages, population pressures and the frequent conflicts in that continent, is the subject of the third chapter. The last paper examines the role of health care for the improvement of conditions in the underdeveloped countries.

Scientists participating in Pugwash meetings are invited personally and represent nobody but themselves. Since the chapters from the individual authors present their personal views, a divergence of opinion on some specific issues is bound to occur, and this is sometimes reflected in the conclusions arrived at by them. It is the task of the Pugwash Council to gather the main drift of the debate and formulate consensus where it is reached. This takes the form of a public statement issued at the end of a Pugwash Conference. The text of the statement published after the Dagomys Conference is given in the Appendix, together with its annex, the Dagomys Declaration "Ensuring the Survival of Civilization".

We believe that the contributions by expert scholars assembled here provide a thought-provoking survey of the most vital issues confronting society at the present time. We hope that this volume will help to elucidate these issues, contribute to their wider understanding, and thus to the building of a more secure world.

In the editing of this book we were greatly helped by Mr. Maxwell Bruce, Dr. David Carlton and Professor Patricia Lindop, to whom we express our gratitude.

We also wish to give our thanks to Ms. Jean Egerton, Mrs. Sue England, Mrs. Natalia Summers and Mrs. Edith Salt for their technical assistance, skill and devotion.

London, Moscow
August 1989

Joseph Rotblat
Vitalii Goldanskii

Contents

Part VII Global Environmental Problems

Part VIII Alleviating Underdevelopment

Appendix

Part I

Strategic Nuclear Disarmament

Deep Cuts in Strategic Nuclear Weapons: Possible? Desirable?

Richard Garwin

Introduction

Will we really be able to reduce the threat of nuclear weapons through arms control? The lustre of the unattainable has dimmed as the goal has come within our grasp; not only are voices heard in the United States on the undesirability of deep cuts, but the current declaratory positions of the governments of France, Great Britain, and China give little encouragement to the United States and the Soviet Union in negotiating further reductions in their own stocks of nuclear weapons.

Are the USA and USSR really more secure facing the 20 000 to 30 000 nuclear weapons of the other side, of which some 10 000 on each side are **strategic** nuclear weapons, with their primary targets within the territory of the other side? Is the rest of the world more secure with 50 000 nuclear weapons in the armories of the superpowers, and some 1000 in the hands of France, the UK, and China (with prospects for acquiring many more)? Or would we be safer with 50 per cent, or 25 per cent, or 5 per cent of these numbers? "Deep Cuts" means cuts much deeper than 50 per cent reductions - at least 75 per cent, or 90 per cent or 95 per cent. In discussing such levels of nuclear weapons, I do not mean to preclude the eventual elimination of nuclear weapons, but the arguments and mechanisms of the present paper cannot be used to support the total elimination. Thus, this is very much an **interim** proposal.

Without repeating my own paper from the Gmunden Conference a year ago[1], I will discuss prospects for reducing strategic nuclear weapons in the light of the events of the last year.

In early December 1987, President Reagan and General Secretary Gorbachev met in Washington. They signed the INF Treaty, eliminating worldwide nuclear delivery vehicles of intermediate and shorter range (500-5500 km), among them the SS-20 Soviet 3-warhead ballistic missile, and the American Pershing-II and the ground-launched cruise missile (GLCM) of which 572 were deployed (or about to be deployed) in Europe.

Indeed, the INF Treaty mandated also the elimination of US and Soviet delivery vehicles of 500-5500 km range for **non-nuclear** warheads, including chemical and high explosives. However, the Treaty put no restriction at all on aircraft or on air-launched cruise missiles or air-launched ballistic missiles, nor did it touch sea-launched cruise missiles (SLCM), of which modern versions, both nuclear-armed and conventional, are being widely deployed in US and other navies. Nevertheless, the INF Treaty will eliminate 283 launchers and 867 missiles from the US side, and 851 launchers and 1836 missiles from the Soviet side, without, however, destroying a single nuclear warhead.

At the Washington Summit, Reagan and Gorbachev instructed their negotiating teams to draft a formal agreement that would reduce strategic weapons by 50 per cent, and promised that nuclear-armed SLCM would be limited separately, outside the Strategic Arms Reduction Treaty (START).

At the Moscow Summit early June 1988, the INF Treaty was ratified by the leaders, but no START agreement was signed. The INF Treaty having entered into force, in 3 years there will be no US or Soviet INF or shorter range ballistic or land-based cruise missiles worldwide. The mechanism introduced to verify the INF Treaty is in the United States lodged in the Department of Defense, and is funded at some $200 million for each of the next three years, presumably with decreasing funding requirements beyond that time. In fact, some of the most extreme verification demands of the United States were accepted by the Soviet Union, and then withdrawn by the USA, apparently because the USA decided that on balance the additional confidence in detecting a violation of the INF by the Soviet Union was not worth either the cost to the USA or the exposure of potentially sensitive US facilities to Soviet verification authorities.

But if the Moscow summit brought no agreement on START, it did solidify the personal relationship between Reagan and Gorbachev. To President Reagan, the Soviet Union is no longer the "evil empire", as he had characterized it a few short years ago. In fact, there was just not enough time to do a responsible job of negotiating a START agreement before the Moscow Summit. In addition, there were critical points of difference remaining between US and Soviet positions, with the Soviet Union demanding strict adherence to the ABM Treaty as it was signed in 1972, with the limits that would impose on testing and deployment of a space-based defensive system (SDI); almost concealed from public view was the US insistence that a period of non-withdrawal from the ABM Treaty would be followed by freedom to deploy ABM defences. There was also the truly difficult problem of controlling nuclear-armed SLCMs, without which constraints the limitation of the frankly strategic delivery systems - ICBMs, SLBMs, and heavy bombers - would be a hollow achievement. Finally, although it is clear that the extent of reductions in US and Soviet strategic nuclear weaponry depends on the future forces of existing and potential nuclear powers, there has been no indication as to how the superpowers view, or would handle, that problem.

In recent months, in a substantial number of publications, experienced, responsible people in the United States and the Soviet Union have touched on the future of strategic forces[2,3,4,5,6,7]. I shall refer to these in passing.

Discussion of Various Professed Positions

First, there is concern expressed for the survivability of the retaliatory force in the largely US-USSR confrontation. For the USA, this is generally taken to require a continued emphasis on submarine-deployed strategic ballistic missiles, and (if land-based missiles are to be retained), a force of mobile, single-warhead missiles. Although many American commentators believe that a fleet of single-warhead mobile missiles would be a suitable retaliatory force for the Soviet Union as well, others emphasize the difficulty in verifying Soviet compliance with a numerical limit, and propose that mobile ICBMs be banned. This, of course, contrasts with the statements of some of those same officials a few years ago, that the US purpose in deploying the MX missile was to "force the Soviet Union to a more survivable basing posture" for its nuclear retaliatory force!

As for the Soviet view, the Committee of Soviet Scientists for Peace, Against the Nuclear Threat, has published several versions of "Strategic Stability Under the Conditions of Radical Nuclear Arms Reductions". Although the Soviet group mentions the ultimate "universal and complete elimination" of nuclear weapons, it does not provide a proposal to achieve that goal except, "...to implement that last stage of nuclear disarmament in a comparatively short period as a single act". Security for all nations and for all individuals in a world totally free of nuclear weapons is a goal at once appealing and equitable. However, I personally do not see how the elimination of nuclear weapons (once achieved) could guarantee the continued non-existence of nuclear weapons; and their resurgence could provide conditions more dangerous to security than the present enormous excess of nuclear weaponry. I will concentrate, therefore, on the (perhaps interim) situation in which the United States and the Soviet Union have on the order of 1000 warheads each.

A.A. Kokoshin[8] has put forth the current views of the Committee of Soviet Scientists, with which I very largely agree. My own position[9] derives from my paper for the 37th Pugwash Conference. The Soviet study emphasizes strategic stability, with minimal nuclear forces that "...should provide only the capability to deliver a retaliatory strike that would inflict unacceptable damage on the attacker. This means that such forces on both sides must be sufficiently invulnerable and reliable".

Significantly, in the Soviet examination of options beyond 75 per cent reductions in US and Soviet strategic nuclear forces, "Calculations show that in order to maintain a stable strategic balance it is necessary for the other nuclear powers to reduce their strategic nuclear forces approximately proportionately to the reductions of those of the United States and the Soviet Union, and at the same time to make similar qualitative changes in the structure of their forces". Indeed, successive versions of the Soviet study at the 600 warhead level proposed a 600-missile mobile force of single-warhead ICBMs, then a 600-silo deployment of similar missiles, and now a mixture of silo-based and mobile missiles. Facing only a US force, the Soviets would (in my opinion) prefer a silo-based force, with its better command and control, and better protection against electromagnetic disturbance; silo basing would be stable against a comparable number of warheads on the US side, however the US warheads are to be based. But if the Soviet ICBMs must survive under threat from US warheads, **plus** an uncertain number from other nuclear powers, the Soviets appear to prefer the survivability of a mobile ICBM.

Drawing on the report of the Commission on Integrated Long-term Strategy[6], US Secretary of Defense Frank C. Carlucci in a speech prepared for delivery at the 25th Wehrkunde Conference in Munich 02/07/88 wrote, "In the next decade, Soviet and American missiles can be made accurate enough to destroy most military targets with a conventional warhead - all but those that are deep underground, super-hard, or of unknown location".

In fact, this statement can readily be misinterpreted. Only if the Soviet Union were to **cooperate** in the non-nuclear destruction of its steel-reinforced concrete missile silos, would the United States be able to pose such a threat. Whereas a nuclear weapon will destroy a silo from any distance less than 100 metres, a conventional weapon carried on a strategic ballistic missile would have to strike the silo cover itself. It could be defeated by a substantial tent over the silo, to hide the exact location. A non-nuclear strategic weapon in the form of a "long-rod penetrator" can have its effectiveness eliminated by mounting on the silo cover itself a large-scale analogue of the "reactive armour" used on modern tanks (Soviet, American, Israeli, German). American technologists know this, as do Soviets.

US security forces need not fear the destruction of US ICBM silos by Soviet non-nuclear strategic weapons; we can **defend** against such. Indeed, we have long had designs for terminal defence (of numerous hardened silos - **not** of soft cities or relatively few hardened command centres) that would work quite well, even against nuclear attack. Furthermore, the cost of a nuclear warhead (even the delivered cost) is small compared with the cost of the silo and the large missile with multiple warheads that it contains, so we must search beyond questions of cost for the desire to have a **non-nuclear** capability to destroy the opposing retaliatory force.

The answer may lie in the understandable reluctance to contemplate the use of nuclear weapons. To quote from "Discriminate Deterrence",

> "The dominant role of the extreme contingencies became especially perverse when they were packed together in public debates with certain other ideas about nuclear deterrence. These ideas became influential in the West by the end of the 1960s. The core idea: that nuclear forces of the United States and the Soviet Union could be locked into a relationship of 'stability' in which 'mutual deterrence' made any war between the two sides quite impossible. There could be no war because war would inevitably lead to the use of nuclear weapons, which would mean the destruction of the two sides, which meant in turn that neither side would ever start a war. Nuclear weapons were inherently unusable.
>
> This doctrine presents large difficulties. If deterrence really depended on mutual vulnerability, then NATO's foundation idea - that an attack on one is an attack on all - would be overboard. In the long run, the doctrine could not even deter selective attacks on the United States. It would be seen as a bluff, and the bluff would be called.
>
> The criticism above is scarcely original. Extreme versions of the doctrine of mutal vulnerability as a guarantor of 'stability' have been assailed for their contradictions ever since they first surfaced. Yet such views have, incredibly, retained an extraordinary hold over political and military elites in the West, especially in Europe".

The third paragraph calls for yet one more resolution of the supposed paradox. It is simply that deterrence depends upon **commitment.** If one does not take care to build systems that are reasonably survivable, there is not much deterrence. If one does not **care** enough to respond to an attack, there is no deterrence. If one **fears** to jeopardize one's own survival, there is no deterrence of anything less than one's own destruction. Of course, speaking in the abstract, deterrence of the Soviet Union depends upon vulnerability of the Soviet Union. Vulnerability of the United States is not at all required for deterrence of a **Soviet** attack. But unless the Soviets were irresponsibly confident that the United States or its allies would never attack, the Soviets will take those measures required to hold the US vulnerable, and hence the Soviets will be confident that the USA is deterred from an attack that it probably would not undertake in any case.

But how about the proposition that if the United States were vulnerable to destruction by a large Soviet attack, then deterrence "...could not even deter selective attacks on the United States. It would be seen as a bluff, and the bluff would be called"?

In a 1977 book[10], I explain that an unexamined assumption underlies many "analyses" of the inadequacy of deterrence. According to such analyses, in case of a massive nuclear attack on a nation possessing survivable nuclear forces, the launching of the retaliatory force would only provoke a further attack on the population and the industry of the victim; thus, retaliation would be withheld and replaced by surrender. Foreseeing this, a would-be aggressor would not be deterred from making the first nuclear attack. Furthermore, he might even see that only a nuclear demonstration would be required to compel surrender, since the logic would be clear to both sides.

The unexamined hypothesis lies in the tacit assumption that the very best decisions are made in every case by having at each moment all available information, considering the results of every course of action, and choosing the best. According to this decision procedure, one would never decide in advance to retaliate under certain circumstances, but would withhold that decision until assessment of the situation after the attack. Then the conclusion follows, and nuclear weapons would indeed not deter a rational aggressor striking a rational victim.

However, if the decision to retaliate if attacked is made **in advance**, the supposed victim is no less rational (rather, uses a non-local decision procedure), and a rational aggressor would indeed be deterred, since he would have nothing to gain (and a lot to lose) from the contemplated attack.

So "Discriminate Deterrence" sets up a straw man.

One may also ask, if these views have "retained an extraordinary hold over political and military elites in the West", why do we have any significant conventional forces on which NATO spends some $300 billion each year? Only because the "elites" have no influence?

Indeed, although it is possible in principle to deter even limited attack by the single possibility for massive destruction of the aggressor, it is probably more effective to deter such an attack by the availability of a more measured retaliation. Thus, as one who has counted strongly on the effectiveness of deterrence by promised retaliation with nuclear weapons, I have not imagined that the sole US capability should be to destroy the Soviet Union, but rather that "selective attacks on the US" would be answered by selective attacks on the Soviet Union. However, I would not waste a few nuclear warheads on destroying a few of the many hundreds of identical Soviet ICBM silos, but would use the first retaliatory nuclear warheads to cause as much damage as possible to the class of targets against which they are directed.

Approaches to Reducing the Strategic Nuclear Threat

It is understandable to want to protect one's nation and one's allies by nullifying the opponent's offensive nuclear strike force:

- by preemptive strike capability,
- by a capability for responsive "war fighting", including the capability to destroy the opposing nuclear force after war starts,
- by **defence**, to render the nuclear weapons of the other side "impotent and obsolete,"
- by non-nuclear strategic offensive weapons, combined with arms reductions, or
- by persuasion, propaganda, lobbying, to persuade the other side by its executive and legislative processes to diminish or eliminate its nuclear strike forces.

A theme underlying the more reasoned and analytical papers is that reductions of 50 per cent in strategic nuclear weapons are acceptable, because rather arbitrary required "coverage" of important military targets can be maintained with the fewer weapons. Although the most important individual military targets have relative worths that may follow a kind of Zipf's law, the nature of conflict emphasizes survivability of the forces, so that beyond a limited number of nuclear weapons assigned to military seaport facilities, major command headquarters, and the like, one comes fairly soon in the scale of values to contemplating as targets many hundreds of equivalent ICBM silos, a few hundreds of similar army division elements, etc.

Therefore, if it were guaranteed that a retaliatory deterrent force totalling a few hundred warheads would be used only against military forces, one could imagine restructuring to survive that blow. With that restructuring, deterrence would vanish. Such great reductions in retaliatory nuclear forces would therefore have to be reversible; or the targets would necessarily not be limited to military targets; or one would need accompanying negotiated agreements to keep the military forces vulnerable to a number of weapons substantially smaller than the number now available. Not an easy negotiation, and not one that I would advocate.

Even with the present 10 000+ strategic weapons on each side, confidence is impaired by the thought that the numbers are still too small because nuclear weapons targets of very little value exceed the number of warheads available. If deterrence is based on the potential destruction of the most valuable targets, it is evident that the warheads 1 up to 100 contribute far more to deterrence than do warheads 9001-10 000, and probably more than numbers 1001-10 000. This is particularly true if the reduction in strategic retaliatory force is accompanied by a lessened likelihood of its destruction before launch, because of comparable reductions on the other side.

Deterrence Does Not Solve All Security Problems

A small nuclear retaliatory force does not solve all problems of national security. For instance, it would do nothing to regain the 52 American hostages held by the "students" in Tehran during the administration of President Carter. But the current much larger nuclear forces do not address that problem either.

As an aid to contemplating a world without the responsiblity induced by promise of retaliation or punishment, one might imagine an urban society without law, in which strictly defensive measures would replace the police, the courts, and the penitentiaries. Instead of driving a small car to work, one might make one's way in an armoured tank, but with a good deal less confidence than now.

Are deep cuts in nuclear weapons on the path to escape from mutual vulnerability? Not so long as there is potential conflict. Although deep cuts make defence against ICBMs more feasible, in the sense that a smaller defence would to the job and a defence of greater fractional leakiness would bring the residual threat to a particular level, it is just this fact that makes defence more and more hazardous as the number of offensive weapons is reduced.

In a deep-cut regime, one cannot seriously and responsibly allow significant defences that might render a small retaliatory nuclear force ineffective. One cannot begin with the 1972 US-USSR ABM Treaty of indefinite duration, and suggest that developments in defence technology should be accommodated by modification of the ABM Treaty to permit some defence, at the same time that one pursues deep cuts in the offensive forces. What was acceptable in 1972 (with no limit on MIRVs to penetrate and destroy ABM systems that might be deployed in violation of the Treaty) is no longer acceptable if we are to have massive reductions in nuclear weapons.

The 1972 SALT agreement limited the USA, for instance, to 1054 fixed land-based ICBMs and to 710 SLBMs, but with no limit on the number of warheads on missiles in those launchers. So a serious START regime is no occasion to allow defences beyond those authorized by the 1972 ABM Treaty. Indeed, it would be useful to have a new Protocol under the 1972 ABM Treaty to further restrict ABM deployment from the present single permitted site to zero.

Space-Based Defences

In his Congressional testimony (17 May 1988), Robert C. McFarlane, Deputy National Security Advisor to President Reagan at the time of the Star Wars speech of 23 March 1983, recounts frankly and clearly the origins of the SDI system. He explains how he and the Joint Chiefs of Staff advocating a greater emphasis on defence **per se** of the Minuteman silos - not of society - and he recalls explaining to President Reagan that the speech the President proposed to make was not supported by the Joint Chiefs or by the technical status of the field. He further testifies that the US strategic defence programme is not directed towards the perfect defence of US and allied societies supposedly called for by the Reagan speech and repeated descriptions by former Secretary of Defense Caspar W. Weinberger, but rather towards defence of the ICBMs, towards improved survivability of the SLBM force by quieting US nuclear submarines, and improved penetrativity of the strategic bomber force by adoption of stealth technology!

In recent weeks, the US press has been full of documentation regarding exaggerated claims made for the status of the nuclear-weapons-pumped X-ray laser as a defence against strategic ballistic missiles. Edward Teller in the 1983-1984 era had claimed in secret letters (now publicly available) that a single X-ray laser module could destroy the entire Soviet ICBM force, if launched simultaneously, and that 100 000 separately directed X-ray laser beams could be obtained. In fact, authoritative statements from the Lawrence Livermore National Laboratory, where the X-ray laser is under investigation, indicate that the X-ray laser in 1988 is **still** not ready for engineering development. In any case, Teller has long claimed that defensive weapons (including X-ray lasers) deployed in orbit are vulnerable to destruction by the other side, and it seems clear that the X-ray laser itself would be a far better weapon to aid penetration by an ICBM force than to use as a defence against it.

The 1987 Report of the Directed Energy Weapons ("DEW") Study Panel of the American Physical Society showed authoritatively and scientifically how far we still are from possessing DEW that could even be **evaluated** for use in an ABM system. The study concluded that it would be at least a decade until a decision could be made whether such DEW have a role in defence against ballistic missiles. One of the most extreme proponents of DEW, Lowell Wood[11], has now fixed upon orbiting kinetic kill vehicles (KKV), dubbed "brilliant pebbles", as the solution to the defence problem.

Kinetic kill vehicles would destroy their prey by **colliding** with them at earth-orbital speeds - at which each gramme of interceptor will deliver as much kinetic energy in the collision as is liberated in the explosion of 8 grammes of high explosive. KKV can be launched from the ground for mid-course intercept or based in orbit as space-based interceptors ("SBI"). Wood ascribes to small SBIs the effectiveness sought from DEW decades hence, but assumes these "brilliant pebbles" could be as quick as ground-based ABM interceptors. But it is wrong to imagine that this is a novel idea, with its flaws, if any, remaining to be discovered. For instance, the micro-KKVs can be countered by local hiding of the targets, even of boosters; and their job can be rendered impossible by the natural evolution of the opposing ICBM fleet to faster-burning single-warhead boosters that might take only 70 seconds to reach full speed of 7 km/s, and would end the very visible boost phase at altitudes far below those at which small SBI could orbit. I have prepared[12] a compilation of seven quotations from my own publications of 1981-1987 proposing such small SBI ("hornets") and identifying effective countermeasures to them.

Particularly Dangerous Nuclear Weapons?

Recurrent official antipathy to particular offensive nuclear weapons (e.g., accurate, 10-warhead silo-based ICBMs) seems limited to those on the other side, i.e., the Soviet SS-18 missile and not the more accurate US MX missile. In reality, there is absolutely no reason to believe that 3000 warheads on 300 large SS-18 missiles pose a greater threat than do the same 3000 warheads on 3000 single-warhead missiles. Furthermore, it is not even true that elimination of strategic ballistic missiles (ICBMs and SLBMs) by agreement, or by deployment of an effective defence, would necessarily give more time for decision and crisis management after the nuclear weapons (bombers and SLCMs) were launched. Surely, if retaliation could be prevented by a zero-warning nuclear explosion in the capital, that could be arranged by detonation of pre-emplaced nuclear weapons whether or not SS-18 missiles exist; therefore, nations must arrange their command and control systems to ensure that retaliation cannot be prevented by sudden attack on the strategic command systems.

Unlimited SLCMs?

Modern nuclear-armed cruise missiles (e.g., the US Tomahawk) are very similar to their conventionally armed counterparts, although the lighter nuclear warhead allows a large fuel tank and a range of 2500+ km in contrast with the 500-1000 km range of the conventional weapon. The submarine-launched Tomahawk is fired from a container loaded like a torpedo into a 21-inch diameter torpedo tube, and is very similiar to versions launched from ships, land (GLCM), and air (ALCM). It has a warhead explosive power more than 10 times that of the atomic bombs that destroyed Hiroshima and Nagasaki.

Except for space-launch vehicles, ICBMs have no non-nuclear or non-military counterparts, so their limitation and verification is relatively simple. Adding to the difficulty of **eliminating** nuclear-armed SLCM by the banning of all SLCM is the perceived value of the conventional SLCM; indeed, proposals have been made in the absence of nuclear-armed SLCM to build a force of hundreds of thousands of non-nuclear SLCM. A further difficulty is the identical appearance of nuclear and non-nuclear SLCM, coupled with the small size and unobtrusive and flexible flight testing of SLCMs. There are, by now, several analyses and reports on options for limiting and verifying limits on SLCM,[13] but they give little support to the thought that there is no need to limit nuclear-armed SLCM in an era of deep cuts in other strategic weapons.

Not only must compliance with treaty limits be verified, but **break-out** must be rendered difficult and unattractive by limiting the number of nuclear warheads available for mounting on conventional cruise missiles. Even in connection with ICBMs, SLBMs, heavy bombers and ALCMs, the US House Armed Services Committee[7] has emphasized the implications of rapid deployment of additional nuclear weapons after a possible denunciation of a START agreement:

Strategic Weapons

Current Forces		Public Perception of START		Adminstration's Aspirations		Potential Outcome with Soviet Breakout	
US	Soviet	US	Soviet	US	Soviet	US	Soviet
12 000	11 000	6000	6000	11 000	9000	11 000	21 000

The Panel observes that the Reagan Administration's force structure planning seems clearly out of step with its negotiating strategy, and I agree with the judgement; the USA should abandon the MX missile and should reverse the trend towards larger but fewer submarines for basing its SLBMs.

What About The Three Other Nuclear Weapons States?

The United Kingdom, France, and China each possess several hundred nuclear weapons. Ultimately, the willingness of the USA and the USSR to cut their nuclear forces depends on their view of the future forces of other nations. While these three nations could not persuade the USA and the USSR to reduce (from the present 25 000) to 1000 warheads each, they could certainly **prevent** such reductions, as discussed in my Gmunden paper[1]. During the last year, statements from Ministers of Defence or of Foreign Affairs of these nations have been uniformly discouraging. Both Andre Giraud, former Defence Minister of France, and George Younger, his British counterpart, have stated clearly that 50 per cent reductions in US and Soviet strategic weapons would have no effect on French or UK plans for modernization of their nuclear forces.

In a speech to the United Nations, 2 June 1988, Qian Qi-chen, Chinese Minister of Foreign Affairs, summarized the position of his government in 8 points, including the reduction of all types of armaments; an ultimate goal of prohibition and thorough destruction of all nuclear weapons; no-first-use of nuclear weapons; complete prohibition of chemical weapons and of space weapons. As regards Chinese plans for their own nuclear forces, Mr. Qian proposed essentially the "three stops and one reduction":

> "The two superpowers that bear a special responsibility should take the lead in putting an end to the testing, manufacturing and deploying of all types of nuclear weapons and in drastically reducing and eliminating all types of nuclear weapons each of them has deployed in any region at home and abroad. Then a broadly represented international conference on nuclear disarmament can be convened with the participation of all nuclear states to discuss what steps and measures to be taken for a thorough destruction of nuclear armaments".

At least since 1984, the position of the Chinese government had been that substantial reductions in US and Soviet strategic weapons would be followed by reductions of Chinese nuclear weapons. More recently, the position appeared to be that a 50 per cent reduction in **all** US and Soviet nuclear weapons would lead not to parallel reductions in Chinese nuclear weapons as the USA and USSR reduced further beyond that point, but to a call by the Chinese for a universal conference to discuss nuclear disarmament. The June 2 speech appears to delay even that minimal action, unless one understands "...take the lead in putting an end to ..." as 50 per cent reductions. In any case, the Chinese, French, and British governments seem to ignore both the promise and the peril of the present situation; in an apparent desire to avoid a future 5:1 inequality, they may be forcing preservation of the present situation in which US and Soviet forces are almost a factor 50 larger than their own.

If the opportunity is missed for START reductions, it is certainly possible that the ABM Treaty will be abandoned, and the nuclear weapons of these three nations will then be faced with the further problem of penetrating defences limited by no treaty at all.

What To Do?

I reproduce here the "deep-cut regime" included in my paper for the 37th Pugwash Conference[1]:

1. The 50 000 warheads total in US plus Soviet inventories would be reduced to 1000 on each side, which would be based on national territory or as SLBMs.

2. Both sides would respect the 1972 ABM Treaty, which would evolve to a total ban on defence against strategic ballistic missiles.

3. Anti-satellite tests would be banned, as would the test and deployment of space weapons of any kind.

   It is important to note that not a single new missile or submarine need be built to move all the way to 1000 warheads on each side[14,15]. Because of the precedent set by the verification provisions of the INF Treaty, one can reduce the strategic weaponry with guaranteed improvement in strategic stability simply by de-MIRVing. Removing all but one warhead from each ICBM and SLBM and leaving only two bombs on each of the bombers would bring each side to a force of 200 warheads. The removed warheads would be demilitarized and the fissile material transferred to IAEA controlled inventory.

4. The nuclear warheads of the United Kingdom, France, and China would be reduced to 200 each.

5. A stringent anti-proliferation regime would be mounted to ensure that additional countries do not acquire nuclear weapons. Part of this regime would be a total ban on nuclear explosion testing, which might be approached by a declining annual quota of low-yield tests.

6. Adequate verification, including cooperative means and on-site inspection as agreed and necessary, would be essential to achieve and maintain this regime.

7. Each side would deploy its permitted weapons in such a fashion as to strengthen stability against first strike. Although each side would explain and open for verification the basing of its 1000 warheads, neither side could compel a particular basing on the other side. In my view, US weapons should be based as follows:

- 400 warheads in the form of single-warhead small ICBMs in soft silos.
- 400 warheads divided among 50 small submarines, each carrying 8 small single-warhead SLBMs.
- 200 warheads carried on 100 aircraft as air-launched cruise missiles, two to an aircraft.

Each warhead must weigh less than 300 kg, including re-entry vehicle for a strategic ballistic missile; the warhead maximum yield would be constrained by the weight limit to about 0.5 megaton. It would be permitted to have any yield lower than that, including the possi bility of variable yield weapons, such as exist at least in the US inventory.

In addition to these points, I propose to ban nuclear warheads from all naval vessels except those dedicated to nuclear-armed SLBMs. All US and Soviet cruise missile production and maintenance facilities would be declared and adequate perimeter and portal monitoring provided by the other side ("PPM") to verify that cruise missiles are fitted with tags and seals to ensure that they have not been equipped with nuclear weapons in the field. All facilities for manufacturing or assembling nuclear weapons would also be declared and monitored, with each nuclear weapon fitted with a unique tag. Clandestine missiles or weapons prepared for breakout could never emerge from their hidden warehouse or cavern, and any such effort would be vulnerable to revelation by one of the staff associated with the activity. I understand that Soviet law dating from 1978 makes duly ratified treaties the domestic law of the land; arms control agreements should include the requirement to publish the agreements widely in the country concerned, and to obtain annual affirmation by those working in the field that their conduct is consistent with the obligations assumed under the treaty, and that they know of no violations of these obligations.

Deep cuts in nuclear weaponry will not happen automatically. Scientists and informed citizens in every nation of the world should make it a priority to encourage and facilitate such reductions and the rule of law and mutual respect under which it can occur. The other nuclear-weapon powers have a special opportunity and obligation to look at their own long-term interests and to take those explicit measures of restraint that will encourage reductions in nuclear weaponry that will benefit them, as well as benefiting the USA, the USSR, and the non-nuclear nations as well.

References

1. R.L. Garwin, "Practical Prescriptions for Deep Cuts", Proceedings of the 37th Pugwash Conference, Gmunden, Austria, 1987, pp. 267-80

2. M.M. May, G.F. Bing, J.D. Steinbruner, "Strategic Arsenals After START: The Implications of Deep Cuts", International Security, Summer 1988, pp. 90-133.

3. S. Fetter and T. Garwin, "Using Tags to Monitor Numerical Limits on Weapons in Arms Control Agreements", prepared for The Project on Technology and the Limitation of International Conflict, The Johns Hopkins University Foreign Policy Institute.

4. "Strategic Survey 1987-1988", IISS, London.

5. S.D. Drell and T.H. Johnson, "Managing Strategic Weapons", Foreign Affairs, Summer 1988, pp. 1027-43.

6. "Discriminate Deterrence", Report on the Commission on Integrated Long-Term Strategy, Co-Chairman F.C. Ikle and A. Wohlstetter, Washington, DC: U.S. Government Printing Office, January 1988.

7. "Breakout, Verification and Force Structure: Dealing With the Full Implications of START", Report of the Defense Policy Panel, House Armed Services Committee, 24 May 1988.

8. A.A. Kokoshin, "A Soviet view on radical weapons cuts", Bulletin of the Atomic Scientists, March 1988, pp. 14-17.

9. R.L. Garwin, "A blueprint for radical weapons cuts", Bulletin of Atomic Scientists, March 1988, pp. 10-13.

10. R.L. Garwin, "Reducing Dependence on Nuclear Weapons: A Second Nuclear Regime", in the 1980s Project/Council on Foreign Relations' book Nuclear Weapons and World Politics, New York: McGraw-Hill Book Company, 1977.

11. "Brilliant Pebbles" Missile Defence Concept Advocated by Livermore Scientist", in Aviation Week & Space Technology, 13 June 1988, p. 151-55.

12. R.L. Garwin, "Hornet history, 1981-1987" (unpublished).

13. For instance, "Potential Verification Provisions for Long-range, Nuclear-armed Sea-launched Cruise Missiles", Workshop Report, Center for International Security and Arms Control, Stanford University, July 1988.

14. R.L. Garwin, Letter to The Editor, San Francisco Chronicle, re reduction of strategic nuclear weapons, 31 May 1988.

15. R.L. Garwin, "Reduce Strategic Weaponry by Half: Can We? Should We?", Remarks for a Panel at a Hearing of the House Armed Services Committee, Washington, DC, 17 May 1988.

A Comprehensive Test-Ban Verification Regime: Implications of Cooperative Measures in INF and START

Jeremy Leggett

Introduction

The verification procedures contained within the INF Treaty have important implications for the relationship between the superpowers, setting as they do the scene for future arms control measures. This paper examines the extent to which INF verification measures can be built on in the negotiation of a START treaty and in the Nuclear Test Ban negotiations. Successful negotiation of a START verification package, especially one capable of addressing the question of mobile missiles (SS-25, SS-24, Midgetman and long-range sea launched cruise missiles), will involve extensive collaboration between the USA and the USSR. I will argue that the kinds of co-operative measure which would have to be involved in a viable START treaty could in turn set the scene, if coupled with judicious use of confidence-building measures, for an effectively verifiable Comprehensive Test Ban Treaty.

Analysis of INF Verification Regime

The INF treaty involves less then 5 per cent of the superpowers' nuclear stockpiles, and current negotiations over new tactical weapons in NATO, plus SLCM deployments by the USA and USSR may mean the global warhead count is not significantly impacted by the INF reductions. Perhaps the chief impact of the INF treaty, then, comes from the standards its verification regime sets for further arms-control agreements.

The following are the most important precedents.

- The USA and USSR will gain their first technical and political experience in using the systematic application of on-site inspection to supplement the use of National Technical Means (NTM).

- The challenge inspections of declared facilities will be truly short-notice. The inspected side will not know where the inspection is to take place until after the inspection team has arrived in their country. Once the location is revealed, the inspected side must prohibit movement of treaty-limited items within one hour and they must ensure that the inspection team can arrive at the site in question within nine hours.

- In the data-exchange, the Soviet Union is providing the West with information about their military programmes far more extensive than anything they have previously disclosed. Encouragingly, "...the data provided by the Soviet side is, on the whole, within the bounds of our intelligence estimates." (Paul Nitze, testifying to the Senate Foreign Relations Committee, 28 January 1988).

- In the process of portal-monitoring, the two sides will gain experience using the various kinds of sensor which would have to be operated (at many more than one location) by inspectors during verification of a START treaty.

- There are encouraging signs in the degree to which the Soviet side has accommodated US fears over the SS-25. The problem for the USA is that because the SS-25 ICBM uses a similar missile stage to the SS-20 IRM, SS-20 production might be continued at the Votkinsk final assembly plant, and a clandestine IRM force might be deployed. Apart from allowing portal monitoring at Votkinsk, where inspectors will use sensors to check that SS-20s are not leaving disguised as SS-25s, the co-operative measure of opening roofs and moving SS-25s into the open at active deployment sites is a very constructive one.

- The task of NTM data-interpreters was also eased when the rules for transit of treaty-limited systems were drawn up. To be included in the data which must be provided are: the numbers of missiles and launchers moved; the mode of transport; the points, dates and times of departure and arrival; and the locations and time at those locations of transported items at least every four days during transit. Their provision greatly assists the task of those reviewing photoreconnaissance and signals intelligence data, and greatly complicates the task of deploying a clandestinely withheld INF force.

The Reagan administration's track-record in contesting Soviet compliance with arms-control agreements suggests that their requirements of a treaty verification would be the most exacting of any US adminstration. Reagan administration officials speak of "effective verification", meaning that "...we want to be sure that, if the other side moves beyond the limits of the treaty in any militarily-significant way, we would be able to detect such violation in time to respond effectively and thereby deny the other side the benefit of the violation." The administration believes it has done this, by means of "...establishing precise, unambiguous obligations, by creating a network of interrelated constraints and measures that allow careful monitoring and reinforce one another in such a fashion as to raise the cost and difficulty of cheating and the risk of getting caught, and by creating a mechanism for addressing compliance concerns." (Paul Nitze in Congress).

START Verification

Verification Regime Agreed in Principle

At the December 1987 summit, the Joint US - Soviet Summit Statement spelt out the verification package which the two sides believe they can negotiate in principle to their mutual satisfaction. The components are as follows.

1. Data exchanges, to include the number and location of strategic weapon systems, and of facilities for production, final assembly, storage, testing and deployment.

2. Baseline inspections to verify these data.

3. Elimination inspections, as the two sides reduce to the agreed level of 6000 warheads and 1600 delivery vehicles, and particular ballistic missile warhead sublimits (which they have yet to agree).

4. Continuous perimeter and portal monitoring of critical production and support facilites to confirm their output.

5. Short-notice inspections of declared locations, locations where treaty-limited systems remain post-reduction (including deployment areas), and locations where such systems have been located (formerly-declared facilities).

6. Short-notice challenge inspections at locations, not listed in advance, "...where either side considers covert deployment, production, storage or repair of strategic offensive arms could be occurring."

7. Provisions which disallow concealment of information vital to treaty verification from NTMs. These would "...include a ban on telemetry encryption, and would allow for full access to all telemetric information broadcast during missile flight."

8. Measures which enhance the observational powers of photo-reconnaissance satellites, to include "...open displays of treaty-limited items at missile bases, bomber bases, and submarine ports at locations and times chosen by the inspecting party.

Other measures, not numbered as above but described in the body of the joint Summit Statement, had been newly-agreed at the Summit, and have important verification implications. These are:

to proceed from the assumption that each type of ballistic missile is deployed with a particular number of warheads, and a commitment that "procedures will be developed that enable verification of the number of warheads on deployed ballistic missiles of each specific type";

to limit long-range nuclear-armed sea-launched cruise missiles (SLCMs), "... and to seek mutually acceptable and effective methods of verification of such limitations, which could include the employment of National Technical Means, co-operative measures, and on-site inspection."

Among the substantive points which have yet to be agreed (at the time of writing) in the Joint Draft Treaty Text are two final points relevant to the verification regime: the counting rule for air-launched cruise missiles on particular types of heavy bomber, and the role of mobile missiles. Whatever the outcome of the ALCM issue, its verification would fall rather simply under measures described in item 8 above. The mobile missile issue, however, will have a more serious influence on the final shape of the verification regime.

Analysis

Notwithstanding the comprehensive and complex nature of the INF verification regime, the challenge of verifying a START treaty effectively is appreciably greater. This is so because weapons systems would be limited, not banned; infrastructures would not be removed, and continued testing of weapons would be allowed.

The data-exchanges, baseline inspections and elimination inspections would be conducted on the same principles as those for the INF treaty. Despite being more extensive, covering as they would a larger number of weapons and weapons systems, they would be able to draw on precedent in the INF treaty.

The continuous perimeter and portal monitoring of critical production and support facilities will also be able to draw on INF experience, at the Votkinsk and Magna plants. As at Votkinsk, the START perimeter-and-portal monitoring effort will be to gauge the extent and specific nature of missile and warhead output, not the absence of activity. Arguably, then, the problem for START will prove to be one of scale, rather than procedure, by the time the treaty is implemented.

Short-notice inspections of declared and formerly-declared locations would build on INF precedents, but short-notice inspections of locations where treaty-limited systems remain post-reduction (including deployment areas), and short-notice challenge inspections at locations of suspect activity not declared in advance, both enter uncharted waters. These are the most intrusive measures of the draft verification package. On the one hand, it is encouraging that both sides concede the need for such co-operative measures, but on the other, it is this which is proving to be one of the most difficult area for negotiators to finalize detail if the Joint Draft Treaty Text is to be carried to fruition. In view of the likely degree of intrusiveness required by Soviet inspectors in US production facilities (in order to assuage professed fears of clandestine production), and in deployment areas which would have to include US Navy vessels, it will also be the most contested point of debate in the ensuing US ratification hearings.

The provisions which disallow concealment of information vital to treaty verification from NTMs go far beyond those agreed during the SALT negotiations. The agreed complete ban on telemetry encryption during missile flight-tests will eliminate one of the most important compliance problems experienced during SALT verification, where an imprecisely-worded treaty article allowed for partial encryption of telemetry. This issue has been a persistent feature of the Reagan adminstration's much-criticized annual reports to Congress on supposed Soviet treaty violations.

The measures which enhance the observational powers of photo-reconnaissance satellites have some precedents in SALT verification, wherein, for example, covers were left off missile silos as a co-operative measure. Displays of treaty-limited items at missile bases, bomber bases, and submarine ports "...at locations and times chosen by the inspecting party" is without precedent in the SALT regime, however, and - if implemented - appreciably reduce the scope for clandestine hoarding of weaponry significantly above the 6000 warhead limit.

The agreement to define a number of warheads on each deployed missile type should make the issue of sublimits easier to resolve. Critically, the number of warheads agreed for the Trident II missile is 8, and this caters for the US view of strategic stability by allowing a larger number of Trident submarines than would have been the case if the SALT counting rule had been applied. For SALT, warheads on particular missiles were counted according to the maximum number the missile had been flight-tested with. For the Trident missile that number is 12. The "penalty" is that by allowing smaller numbers of warheads than the capacity number, still more intrusive co-operative measures of verification will have to be agreed, including the presence of inspectors aboard US Navy vessels. These inspectors will not have to see below missile nosecones, but will need to be able to hold sensors over open missile tubes to assess the number of warheads in place below the nose cone.

The commitment to limit long-range nuclear-armed SLCMs was a major point of progress in the Summit. Previously, the US side had rejected limitations as unverifiable. The Soviet side, having moved before the summit from their professed desire for a complete ban to a ceiling of 400 per side, presented a case-in-principle for SLCM verification which the USA must have deemed sufficiently implementable to alter their own position in the summit week. One possible method would involve the use of sensors operated aboard ships which carry dual-capable SLCMs, where inspections could be limited to the ship's magazine. Other technologies which could be applied to this problem are mechanical and electronic missile tags to assist in the counting process, and tamper-proof devices akin to permissive action links to be fitted to missile casings, thus ensuring that a missile once counted as a conventionally-armed SLCM could not be fitted clandestinely with a nuclear warhead. The technical options are numerous, but all of them involve an unprecedented degree of intrusiveness. The mobile missile issue, if it is to be solved by limiting (rather than banning) the missiles concerned (SS-25, SS-24, and Midgetman), may require similar use of tags to assist the counting process. However, co-operative measures at deployment sites (for example observable discriminants on rail-based deployments) may go some way towards easing the counting problems.

In conclusion, the agreements-in-principle over verification of the START treaty suggest that such a treaty is negotiable. The technical problems over verifying the measures agreed-in-principle can be overcome, though the negotiation of an unambiguous treaty will be difficult, and the verification infrastructure required enormous.

If the START treaty as currently envisaged comes to fruition - as it can and must - cooperative measures over verification will have a knock-on effect which will take prospects for arms control into totally new territory during 1988 and beyond. The next section examines the implications for verification of a CTBT.

Test-ban Verification

An Office of Technology Assessment report to Congress, the product of an exhaustive technical review process, is due out shortly. It will conclude that cavity decoupling in salt deposits poses the only realistic opportunity for staging a clandestine testing programme. Other potential evasion techniques (which space prohibits from covering here), though not inconsequential, are to be rejected by the OTA as insurable against. The OTA report further concludes that because of the stability of the cavern the evasion gauntlet could only be run up to a maximum of 10 kilotons. The evasion gauntlet, then, is constrained not just by seismological, but also geological and operational considerations. As I demonstrate below, though the seismological considerations are the most formidable, the geological parameters (which constrain cavern location, construction, and stability) and the operational considerations (which constrain concealability) add substantially to the gauntlet in ways which are usually insufficiently aired in the debate over test-ban verifiability.

The Seismological Gauntlet

In the USA, opinion varies as to the maximum yield at which the Soviet Union could explode a bomb in a cavern and stand a chance of not being discovered by a reasonable network of in-country seismic monitoring stations. The published worst-case consensus lies somewhere between 5 and 10 kilotons. Among those who have aired this view are a senior seismologist at the Lawrence Livermore National Laboratory, and its Director (in Testimony to Congress). In the UK, Ministry of Defence view that the Soviets could in theory get away with a clandestine test of "several tens of kilotons" is, therefore, seriously out-of-step with worst-case thinking in the USA. Accordingly, a revision of the documentary basis for the UK Government's view (a July 1985 submission to the UN Conference on Disarmament: document CD/610) would be timely.

More optimistic appraisals are to be found among seismologists outside the weapons laboratories. One detailed analysis, by seismologists from the US Geological Survey and the University of Colorado, concludes that a network of 25 stations in the Soviet Union would be able to discover (that is detect and discriminate) clandestine explosions at all possible sites in the Soviet Union, even if the tests were cavity-decoupled (Evernden, Archambeau and Cranswick, Reviews of Geophysics, 1986).

In the last analysis, though, empirical data about the propagation of seismic energy in the Soviet Union are too few to allow anything other than extrapolation from better-understood areas of kindred geology. The OTA report, in view of this, reserves its judgement on the number, type, and location of seismic stations which might be needed to verify compliance with a 1 kiloton LYTBT treaty. There is no doubt from existing data, however, that the seismological gauntlet faced by a test-ban evader would be substantial. Data being gathered currently by existing civilian-domain monitoring stations, especially outside the Soviet Union in Norway and inside the Soviet Union in Kazakhstan (the three stations operated jointly by the Natural Resources Defense Council and the Soviet Academy of Sciences since May 1986) illustrate the extent of that gauntlet. For example, subkiloton tests from Kazakhstan have been detected and discriminated 4000 km away in Norway, and a 20 ton (i.e. 0.02 kiloton) chemical explosion on the Semipalatinsk test-site has been detected clearly at the distance of 300 km by an NRDC station ...in the midst of the seismic waves from a strong earthquake.

The geological and operational gauntlet

Seismology is not all that constrains the verification gauntlet. In this section I consider the full range of geological and operational parameters in the evasion gauntlet, and argue that they point to avenues for CBMs which can in principle build confidence in CTBT compliance. Geology makes for uncertainty over the degree of decoupling to be expected when a bomb is detonated in a clandestinely-prepared cavern because:

- experimental experience (in the west) is thin, and results have not matched theory;

- salt is both physically and chemically heterogeneous in nature;

- solution-mining cannot fashion a circular cavern.

Further uncertainties would arise in the potential evader's mind because Soviet and American salt deposits are in areas of low ambient seismicity where the monitoring side could concentrate in-country seismograph stations.

Operational considerations require the potential evader to run a photo-reconnaissance and signals intelligence gauntlet of unknown dimensions. With suitable allocation of priorities for his non-seismological NTMs, the verifier can make this gauntlet appreciable. This is so for the following reasons.

1. A military-worthwhile clandestine testing programme must involve communications traffic. (Past precedents of SIGINT detections of preparations to conduct nuclear tests have been compiled by Desmond Ball). Under the CTB the problems for the evader will be particularly acute: all communications would be high risk, since the entire testing infrastructure would be supposedly defunct.

2. The engineering operations for a clandestine testing programme would have to be substantial.

3. The positions of existing large standing caverns are known, so that detailed photo-reconnaissance can be focused on them if suspicions arise.

4. The time required for solution-mining a cavern large enough for a testing programme is substantial. The locations of most or all Soviet solution mining projects, for example, are known to the US Geological Survey. The fact that a covertly-mined cavern would take a minimum of several months to prepare adds to the risk of attracting the verifier's attention.

5. Even if a cavern can be solution-mined without attracting comment, access would have to be mined in order to prepare a test, or inspect the cavern if it were to be reusable. Such operations are not required for normal storage or extraction operations.

6. If the clandestine testing programme were to need a series of tests, experimental precedents and engineering considerations suggest that caverns could not be relied on for reuse. If a number of caverns were to be needed, the gauntlet is compounded.

After testing, quite apart from the seismological uncertainties, the evader would need to be sure of the stability of his cavern. Uncertainties arise from the variable geological properties of salt, and industry experience suggests that the evader would face difficult-to-quantify risks of both collapse (leaving detectable, suspicious, surface effects) and leakage (leaving detectable, though not source-locatable, atmospheric contamination).

Other factors are pertinent:

- Military considerations. Would a SERIES of tests be needed? How big does a test have to be before its military utility makes contemplation of a clandestine testing programme "worthwhile"? Even if a new warhead design is finalized by a clandestine programme, what then?... the new warhead must be produced and deployed before a military advantage is accrued; ... how big is the follow-on evasion gauntlet?.

- Economic and political considerations. Would a renewed arms-race, in the event that the clandestine test programme were discoverd, be overly detrimental to the potential evader's economy? Would the risk of a collapsed Non-Proliferation regime add to the nuclear threats posed to the evader?.

Scope for Confidence-Building Measures

Given that a period of years would be needed to transform a clandestine testing programme into a strategic stockpile advantage, the initial verification regime for a comprehensive or low yield threshold test-ban can be less-than-perfect, as it is for the INF treaty. A less-than-perfect verification regime might consist of a limited number of in-country seismograph stations and a limited number of challenge on-site inspections. However, there are opportunities for building mutual confidence in compliance, as time goes by. Some of these CBMs it would be in the interest of all parties to the treaty to implement, as a hedge against mischievous manipulation by opponents to arms-control of non-compliance allegations intent on undermining confidence in the treaty.

Some of the CBMs may appear utopian, but when viewed in the context of the kind of intrusive verification which will be required for treaties in the areas of strategic arms and conventional forces, they will seem less so if - as seems likely - more progress is made across-the-board in arms control. The potential CBMs include the following.

1. Empirical calibration of seismic monitoring networks by chemical or nuclear explosions.

2. Creation of an inspectorate for chemical explosions above a particular size, pre-announcement of chemical explosions, and making available challenge on-site inspections for such explosions.

3. Monitoring of the dismantling of nuclear testing infrastructure - both at known test sites, and in plants producing specialized engineering apparatus specifically for nuclear testing, drilling equipment, weapons effects-tests tunnel equipment (relevant to a CTB only).

4. Post-treaty "key personnel" monitoring ...data-exchange on personnel who have gained critical field experience of nuclear testing in the past decade, and means of verifying their subsequent employment (CTB only).

5. Register of mining operations as a means of aiding photo-reconnaissance interpretation, and selection of sites for challenge on-site inspections.

6. Creation of open-file "cavity-decoupling risk maps" of areas underlain by thick salt deposits, based on satellite photo-interpretation, using 10m or 5m resolution data of the kind currently commercially available, selectively ground-truthed by reciprocal field inspections and study of seismic reflection profiles.

7. Installation of tamper-proof in-country radiation monitors in areas of high cavity-decoupling risk.

8. Collaborative decoupling experiments.

9. Register of sizes, locations and purposes of all past peaceful nuclear explosions (USSR only).

10. Provision of in-country seismic data for past nuclear tests to the ISC (USSR only).

Soviet-American Collaboration on Seismic Monitoring

Mikhail Gokhberg

This paper reports on two ongoing projects. The first project concerns an international exchange of waveform data. The second project is a collaboration between the Soviet Academy of Sciences and the Natural Resources Defense Council (NRDC), recently joined by a British Group, which will install an additional station.

International Exchange of Waveform Data

The Soviet Union is taking part in an international experiment arranged by the Group of Scientific Experts in Geneva with the purpose of improving the global system for verification of a nuclear test ban. The Institute of Physics of the Earth will act both as a National Data Centre (NDC) and as an International Data Centre (IDC). The Academy of Sciences of the USSR has also become a member of the International Research Institute for Seismology (IRIS). In the framework of the experiment which is under way the participation of the four IDCs - Moscow (USSR), Washington (USA), Stockholm (Sweden) and Canberra (Australia) is envisaged.

Each IDC will accumulate seismic information of the first and second level from the NDCs, formed in the four regions. Each NDC accumulates and initally processes the seismic information from the national stations network. Each IDC, in its turn, after the information from the NDC of its own region is collected, will transmit it to the other three IDCs. Thus, all the four IDCs will have identical information from the stations of the entire world. The first level data describe the seismic signals parameters, and the second level data insert the recording part of the seismic signal prior to the first "arrival" of the seismic noise. The decision was made to insert a 30-second recording before the seismic signal to the second level data, and a 60-second recording after the first arrival of the straight P-wave. All data will be transmitted in digital form.

The following seismic stations in the USSR will participate in the experiment: Obninsk, Arti (Sverdlovsk region), Irkutsk, Garm, and Kislovodsk. All these seismic stations have equipment corresponding to the IRIS standards.

The Centre located in Obninsk will function as the National Data Centre of the USSR. The centre functioning as IDC will be also located there. The computers of PDP 11/84 type and SUN 2 will be used at the NDC and IDC.

The channels of GTN WMO (Global Telecommunication Network of the World Meteorological Organization) are supposed to connect the NDCs of other countries. Satellite channels with the speed of 64 Bt/sec and will be used for communication with other IDCs. At the same time the information from Obninsk will be transmitted via the Soviet satellite "Luch-1" to Lvov and through the "Intelsat" system to other IDC.

Collaboration on Seismic Monitoring

An investigation of the noise and the main seismic wave field properties at Semipalatinsk was performed using three temporary stations: BAY, KKL and KSU. The high-frequency seismic noise at the three stations was measured both at the surface and the boreholes. The borehole sensors were at depths between 66 and 101 metres. The basic parameters of the microseismic noise, the hypocentre location of the main signal sources, their types and the signal attenuation with frequency, have been estimated during the experiment in Kazakstan. In the band range 0.25-1 Hz the microseismic spectrum falls off with the frequency as f^{-4}; at a frequency of 3 Hz (at stations BAY and KSU), the radiation varies as f^{-2}, and at KKL as $f^{-2.5}$. On the surface, the station KKL appeared to be the noisiest (at 2.5-7 Hz it is 1.5-2.5 times larger than in BAY). At the frequency above 10 Hz the noise in the KKL borehole becomes lower than on the surface; this starts for the BAY station from 8.5 Hz, and at KSU from 3 Hz. At high frequencies, for instruments installed on the surface, the wind influenced the noise level, but there is no such dependence for the borehole sensors. Industrial explosions of chemical substances appeared to be the main signal sources in Kazakhstan. Remote and teleseismic earthquakes were recorded as well.

In September 1987 three special chemical explosions were performed. The first and third were near Karaganda and used 10 tons of TNT. The charges were placed in boreholes and at a depth of 20 metres. The second explosion was near the Semipalatinsk test site and used 20 tons of TNT. The charges were placed in a tunnel system. During the explosion a throw-out occurred, resulting in a weakened activity and a noticeable radiation asymmetry. The magnitude estimations of this event varied at different stations. For this event the ratio of the amplitudes of the P and S-waves differed from the other two. A remarkable difference was present in group S itself and in Rayleigh waves. Thus, this characteristic can be used to establish how the charge was destroyed and whether throw-out took place.

The third explosion was recorded at the background of the teleseismic event. Since instruments with different frequency characteristics were used for recording, we managed to get a clear record of the event itself.

All explosions were clearly recorded at all stations. As the longest distance involved was 630 kilometres, one can conclude that very small explosions could be monitored at large distances in the Semipalatinsk region.

A similar experiment was carried out in Nevada, April 29-30 1988. Three chemical explosions were used at Lathrop Wells, Broken Hills and Black Rock Desert. They were recorded at 3 seismo-stations: Deep Spring (DSP), Nelson (NEL) and Troy Canyon (TRC) with the same kind of instrumentation as used in the Semipalatinsk experiment. The charges varied from 10 to 14 tons. The explosions were performed in boreholes to ensure good coupling. The Black Rock Desert explosion was clearly recorded at DSP and TRC, but at NEL (distance 679 km) the explosion signal was lower than the noise level. In Semipalatinsk the signal-to-noise ratio was significantly higher at that distance. The Lathrop Wells explosion was recorded at NEL (distance 159 km) in the borehole site, but the surface site recording had a very low signal-to-noise ratio. The same was true for the recordings at DSP (distance 181 km). The recordings at TRC of the Broken Hills shot were good, but at a distance of 470 km (NEL) the explosion signal was lower than noise level.

The data on the noise at the stations in Nevada were also analysed. The absolute micro-seismic noise level was close to that in Kazakhstan. The TRC station appeared to be the noisiest one. The form of the spectra curves was close to Kazakhstan ones. Among the recorded events, compared to Kazakhstan, the earthquakes, occurring in the bordering regions of California were in the majority. Using the envelope of the coda method the Q-factor can be calculated. This parameter is strongly dependent on frequency at Nevada compared with Kazakhstan. At frequencies from 8 up to 12 Hz the values are of the same order for both regions. At frequencies lower than 8 Hz and higher than 12 Hz, the Q-factor in Semipalatinsk is significantly larger. Using the data on P-wave arrivals and polarization methods the accuracy of hypocentre determination was estimated. An the centre of the test-site it appeared to be about 5 km for sources of 10 tons; beyond the triangle boundaries, at a distance of 100-150 km from the stations, it was 15-20 km. The spectra analysis of P-wave group showed that their slopes at high frequencies noticeably differed from explosions and earthquakes. This peculiarity could be used to identify the type of source.

In addition to the seismic measurements, the ionosphere disturbances caused by the explosions were monitored. The Doppler shift method was used to study the movements of the 100 to 200 km levels. This was done together with a group from Stanford University but the results of the experiment have not yet been analysed.

In an earlier experiment, the dependence of the electronic density vertically above the station of heights between 100 and 700 km was measured as a function of time. The explosion was performed in the Apatity mine in the Cola peninsula. At a distance of 600 km from the explosion, disturbances start to appear when an acoustic wave travelling with the speed of sound reaches the measurement point. They are caused by an acoustic wave effect on the ionosphere. The density of the atmosphere decreases exponentially with height and this causes the amplitude of the acoustic wave to increase exponentially. The signal in the atmosphere is however considerable.

The number of waves with different velocities, generated in the atmosphere by explosions, requires more detailed analysis. Monitoring of ionospheric disturbances will not be the main method for detection and identification of explosions but it could be used as a supplementary method; at present it is used to predict tsunamis.

Research on its use for monitoring of underground nuclear explosions was started in 1980 in a study by Elisabeth Blane on explosions in Muroroa. The work in the Soviet Union started in 1981 using surface explosion in Central Asia. Both ground-based and satellite-based observations were made. The explosion involved 251 tons of TNT and could clearly be seen in the ionosphere and in the magnetosphere.

Threshold Limits on Anti-Missile Systems

John Pike

Introduction

New developments in ballistic missile defence (BMD) technology pose major challenges to the Anti Ballistic Missile (ABM) Treaty of 1972. There is a growing need to find a way to keep the ABM Treaty current with the evolution of BMD technology. While the ABM Treaty is of indefinite duration, it needs periodic updating.

Over the past two years, discussion has focused on the debate over the Reagan administration's so-called "broad" interpretation of the ABM Treaty, which holds that the Treaty does not limit the testing of exotic BMD technologies, such as lasers. But this is a false issue. It is increasingly clear that this interpretation of the Treaty is without legal or factual merit. Congressional opposition seems likely to ensure that it will not be implemented.

Resolution of the broad interpretation debate does not resolve the inevitable conflict between the permissive and restrictive readings of terms of the traditional interpretation of the ABM Treaty.

The Reagan administration espoused a permissive reading of the traditional interpretation of the ABM Treaty in it's first annual Report to the Congress on SDI in early 1985, and in each subsequent edition of this report. The administration asserted that the SDI programme was consistent with this permissive reading of the Treaty, arguing that the SDI was not developing ABM "components," and that SDI tests would not be conducted "in an ABM mode," or demonstrate "ABM capabilities," and thus the technologies tested under SDI would not be "capable of substituting for" ABM components.

But a more restrictive reading of the Treaty's terms leads to the conclusion that many of the tests under SDI do involve components with ABM capabilities, and thus are inconsistent with the Treaty. Unfortunately, the Treaty provides inadequate guidance for choosing the proper reading of these critical terms.

New definitions of what constitutes ABM "capabilities," and focusing on thresholds rather than categorical bans, could resolve this problem. Devices with capabilities above a certain "threshold" would be subject to the testing and deployment limits of the Treaty, while those with inferior capabilities would not. Similarly, there are questions about what is an ABM "component" or what constitutes "development", terms that are central to the ABM Treaty, but which lack sufficiently precise definition. Threshold limits would provide a less ambiguous operational definition for the "development" of an "ABM component" which has "ABM capabilities" or has been "tested in an ABM mode."

This approach was first proposed in a meeting sponsored by FAS with Soviet scientists in early 1984. In 1986 and 1987 a series of private talks confirmed the usefulness of threshold limits.

The Soviet government formally proposed negotiations along these lines in July 1987, and September 1988. While the details of the Soviet proposal have not been made public, it has been reported that the Soviets wanted to ban space-based testing of kinetic kill mechanisms, set ceilings on the power levels of laser kill mechanisms to be tested in space, and regulate testing of high-energy power sources in space.

While senior arms control advisor Paul Nitze has publicly expressed interest in reaching an agreement of this kind, and the Joint Chiefs of staff have privately encouraged this approach, the American government remains committed officially to the broad interpretation of the Treaty.

The ABM Treaty

The ABM Treaty of 1972 is the only bilateral arms limitation agreement in effect between the United States and the Soviet Union. The Treaty reflects their shared judgement that limitations on strategic defences and offenses are interrelated. The Treaty serves two mutually supporting and yet contradictory roles in limiting anti-missile systems.

The first role of the Treaty is to permit research in support of the Treaty regime. Research on BMD technology provides reassurance that BMD systems would not be effective, because both sides understand the technology well enough to design countermeasures. Designing a decoy requires a practical familiarity with the performance characteristics of the sensor that it is designed to fool.

The second purpose of the ABM Treaty is to establish a long lead-time for deploying an anti-missile system. Sufficient time should elapse between the point at which the treaty regime was exceeded, and the time an anti-missile system was actually deployed, that there would be no requirement to implement countermeasures in existing force structures.

The essential problem is finding a balance and a threshold between the low level of research that is necessary to develop effective countermeasures, and thus uphold the ABM Treaty regime, and the high level of activity that would permit a system to be deployed quickly - provoking the deployment of those countermeasures, particularly the proliferation of offensive forces, and thereby impeding negotiated limits on offensive systems.

The Treaty prohibits the development, testing and deployment of ABM components that are space-based, air-based, sea-based or mobile land-based. The Treaty provides several criteria for establishing what devices are subject to these limits:

- the components of ABM systems at the time of the signing of the Treaty, namely interceptors, launchers, and radars;

- devices that have been "tested in an ABM mode" (that is, tested against strategic ballistic missiles or their components in flight trajectory);

- devices that have "ABM capabilities" or are "capable of substituting for" ABM components.

Variant Readings of the ABM Treaty

The Reagan administration has offered two variant readings of the ABM Treaty that both serve to negate these prohibitions.

The so-called "broad" interpretation of the Treaty, holds that the testing limits of Article V do not apply to exotic systems based on other physical principles, and that the only provisions of the Treaty relevant to devices other than conventional rocket and radar systems is the Agreed Statement D limit on deployment of exotic systems. A variant of the broad interpretation of the Treaty holds that even kinetic energy weapons are not subject to the Treaty's testing limits, since such devices have on-board guidance systems, and thus are not totally dependent on external guidance from ABM radars or other sensors.

The recent Senate debate over this issue has made it clear that the broad interpretation is very unlikely to become the basis for the SDI programme. In any case, the administration has had considerable difficulty explaining what additional tests would be conducted should the broad interpretation be implemented.

The Reagan administration's permissive reading of the narrow interpretation of the Treaty recognizes that the Treaty does apply to all types of anti-missile components (including exotic systems). But it holds that the Article V restrictions on mobile components does not contain the SDI, since none of the devices tested under the programme have **all** of the characteristics of an ABM component, and that Article VI restrictions do not constrain SDI testing since these tests would not actually and totally demonstrate ABM capabilities.

The Zenith Star Space-Based Chemical Laser

These problems of Treaty interpretation are best illustrated by the case of the Zenith space-based chemical laser that will be tested in orbit in the early 1990s. The Zenith Star test was originally formulated as one of the four tests that would be conducted under the broad interpretation of the Treaty. When it became clear that implementation of the "broad" interpretation faced insuperable political difficulties, new rationales were developed to support the claim that this test was consistent with a permissive reading of the traditional interpretation of the Treaty. But a more restrictive reading of the traditional interpretation of the ABM Treaty would conclude that this test is inconsistent with the ABM Treaty's Article V ban on space-based components. Under the provisions of Article VI, this would be the case even if the tests were against a satellite target, rather than ballistic missile component targets.

According to an article that appeared in the 30 November 1987 issue of "Aviation Week and Space Technology" there are four lines of reasoning that support the claim that Zenith Star would not be a component, and would be compliant with the Treaty:

1. it will not be tested against strategic ballistic missile targets;

2. it will only have a short range capability;

3. it will engage cooperative targets;

4. it will not have a target acquisition capability.

The first rationale, according to Aviation Week, is that "neither ballistic missiles nor their substitutes will be used by SDI as targets, nor will other vehicles in ballistic trajectories be illuminated by the Alpha laser". This suggests that Zenith Star will be tested against satellite targets. But discussions with contractor personnel suggest that Zenith Star will be tested against ballistic missiles targets.

Article VI of the ABM Treaty provides two criteria for what constitutes a Treaty-accountable component: testing in a ABM mode (against targets with the characteristics of strategic ballistic missiles); and demonstration of ABM capability. While Zenith Star may not be tested in an ABM mode, it will demonstrate ABM capabilities when tested against satellite targets, and is thus prohibited under the Treaty.

The second rationale according to Aviation Week is that "Zenith Star will have only a short range capability ... (against) ... a target 100-200 km away". But the potential brightness of Zenith Star is of the order of 10^{18} watt per steradian. Calculations from first principles suggest that the minimum brightness needed for one laser to intercept a single liquid-fueled booster with a five minute burn time is on the order of 10^{16} watt per steradian. This defines the minimum brightness that could be considered to be ABM capable. Although a system with greater effectiveness might require lasers with a brightness four to seven orders of magnitude greater, the ABM Treaty applies equally to very effective and not-so-effective systems and components. With a brightness 100 times higher than the minimum that defines ABM capable, Zenith Star is thus Treaty accountable under Artice VI even if actual tests are only conducted at short range.

While the effective range of a direct energy weapon may be an appropriate measure of its ABM capability, and this may be a useful topic for future negotiations with the Soviets, effective range is not currently part of the ABM Treaty, and thus this threshold cannot be used in assessing the compliance of a test.

The third rationale, according to Aviation Week, is that "all of the spacecraft's targets would need to be cooperative with Zenith Star and augmented with transponders or laser reflectors to interact with the laser spacecraft".

The 1972 Senate ratification testimony of Ambassador Gerard Smith pointed out that "The fact that early stages of the development process, such as laboratory testing, would pose problems for verification by national technical means is an important consideration in reaching" the definition of permitted and prohibited activities.

Verification using national technical means is the crucial criterion in determining ABM Treaty limits. It would be extremely difficult, if not impossible, to observe transponders or laser reflectors on Zenith Star targets by national technical means. And even if such devices were observed, there would be no reason to believe that these cooperative devices were in fact critical to the success of the test. If the United States were to be presented with such a test by the Soviet Union, surely we would raise questions as to whether the cooperative devices were merely a deception.

The fourth rationale, according to Aviation Week, is that "the spacecraft will require prior programming on where to look for specific test targets and will have no autonomous target search capability".

This seems to assume that a device would not be accountable under the Treaty unless it is simultaneously capable of substituting for both an ABM interceptor and an ABM radar. Even if Zenith Star cannot substitute for an ABM radar, it can substitute for an ABM interceptor, and is thus Treaty-accountable. The Treaty applies not only to ABM systems, that can both track and intercept targets, but also to ABM componenets, that can either track or intercept targets.

The asserted compliance of Zenith Star assumes that a device would not be Treaty-accountable unless it could perform the complete function of or substitute on a "stand alone" basis for an ABM system as defined in Article II of the Treaty. If a device could only perform the function of an ABM radar, launcher, or interceptor, or a part thereof, then it would not be constrained as an ABM component under this reading of the Treaty.

There is no instance to date of a single stand-alone device that is capable of simultaneously substituting for both an ABM interceptor and an ABM radar. Although some missile defence systems have a single sensor, they are the exception, rather than the rule. For example, the early Nike-Zeus system had not one or two, but four separate types of radars, for target acquisition, decoy discrimination, target tracking and interceptor tracking. Under the permissive interpretation of the difference between a "component" and an "adjunct", all of these radars would be considered to be adjuncts to one another, and none of them would be considered to be a component.

The Core Problem

The central problem is that the march of technology has complicated the interpretation of the terms of the Treaty. In 1972, verification of testing in an ABM mode was a fairly straightforward process. The operation of a radar could be monitored by electronic intelligence satellites, and the launching of an interceptor, and the flight of a target reentry vehicle could be monitored by various means. These activities provided a rather unambiguous basis for defining "tested in an ABM mode".

But the new BMD technologies pose a greater challenge for determining whether a device has been "tested in an ABM mode". Passive sensors, such as telescopes which can be used to track targets, do not emit signals, and thus their association with an anti-missile test can be difficult to determine. Long range interceptors can be tested against satellite targets which mimic the characteristics of a strategic ballistic missile.

Unfortunately, the determination of whether a device is capable of substituting for an ABM component or whether it has ABM capabilities is also very difficult, particularly if the device is based on new physical principles. The ABM Treaty does contain a precise threshold definition of what constitutes a radar that has ABM capabilities, but the Treaty provides no guidance on the point at which a tracking telescope is capable of substituting for an ABM radar.

Threshold definitions of ABM capabilities could resolve this problem. There may be questions about what is an ABM "component" or what constitutes "development", but it should be possible to determine whether or not a mirror is larger than two metres in diameter, with an acceptable margin of error. These threshold limits would provide a less ambiguous operational definition for the "development" of an "ABM component" which has "ABM capabilities" or has been "tested in an ABM mode".

New Threshold Limits

A ballistic missile defence system is composed of four elements - weapons, weapon launchers, sensors and battle management. It is generally recognized that battle management poses the greatest technical challenge for perfecting an anti-missile system, and that sensors pose a greater technical challenge than do weapons and weapon launchers.

It is an unfortunate paradox that the most technically challenging aspect of a BMD system (battle management) also poses the greatest challenge to verification, while the least technically challenging part of the problem (weapons) pose the least challenge to verification.

The ABM Treaty places no constraints on battle management systems, since it was recognized by both parties that such limitations would be difficult if not impossible to verify.

At the time the ABM Treaty was signed, BMD sensors were very large radars that required years of construction, and thus were easy to verify. So the Treaty provided a strict regime of limitations on the deployment of such radars. But future systems using passive sensors may be much more difficult to verify. This does not mean that such future sensor technologies would be impossible to verify, or that they should be exempt from constraint. But more stringent constraints on weapons testing may be needed to compensate for the difficulties of limiting sensors.

Criteria for Threshold Limits

First, the threshold should ideally apply to a wide range of technologies. One of the challenges posed by new and emerging anti-missile technologies is their dazzling variety. The search for threshold limits should focus on a small set of parameters that cover a wide range of weapons, sensors, or both. A common limit of five square metres on the aperture of laser beam director mirrors, satellite sensor mirrors, and the windows on air-borne sensor aircraft, would constrain a wide variety of weapon and sensor technologies. The range of systems that could require nuclear power sources is one reason for seeking limits on space reactors.

Second, the limits should apply to technologies that are of interest for ballistic missile defence. While limits on some systems, such as rail-guns, might be imagined, the low priority currently assigned to such devices suggests that more immediate issues, such as lasers (particularly ground-based), rocket interceptors and passive infrared telescope sensors should be addressed as a matter of priority.

Third, the threshold limit should be related as directly as possible to the actual performance of the device in question. The power aperture product, the potential threshold limit that was agreed to in 1972 does this very well. The brightness of a laser is similarly a very good measure of the lasers military performance.

Fourth, it must be possible to distinguish between permitted and prohibited activities. Ten or twenty years ago the volume of an interceptor was a fair indication of its anti-missile potential. But the recent advent of very small terminal homing sensors has reduced the size of interceptor warheads, and thus of interceptor rockets to the point that ABM interceptors are today the same size as anti-aircraft interceptors.

Fifth, the threshold should provide adequate insurance against breakout from the Treaty limit. Limits on deployment provide at most five years lead-time, and may provide much less, while development and testing constraints may provide a five to ten year lead-time.

And sixth, the threshold limit clearly must be verifiable. The required technical collection systems and other means of verification should be available during the time-frame in which the parties to the Treaty are likely to encounter the thresholds they are intended to monitor. The cost of the monitoring system should be less than the cost that would be incurred by not placing a limit on the activity in question. And technical collection systems should not be so capable that they reproduce the anti-missile systems that they are intended to limit. Large space-based infrared telescope sensors used for verification may be difficult to distinguish from sensors that would form the basis for an ABM battle management system. It would be perilously paradoxical if it were necessary to develop or deploy reasonable facsimiles of an anti-missile system to verify limits on the development or deployment of such a system.

New Approaches to Verification

In some cases verification may require the use of non-intrusive co-operative measures. While national technical means may be adequate to measure the aperture of the beam-director mirror of a ground-based laser, when not in use such mirrors are normally screened from the environment by a moveable cab or dome, and thus out of sight. Agreement would have to be reached that such screens would have to be temporarily removed on a periodic basis to permit monitoring by national means.

Some threshold limits might require more intrusive means of verification. In-country monitoring station may be needed for the verification of limits on threshold limits on the brightness of lasers. During its passage through the atmosphere a fraction of the energy of a laser beam will interact with the atmosphere, through such mechanisms as aerosol scattering. An automated collection device, stationed a few kilometres from the laser beam director, could observe this scattering, and determine the laser's wavelength. With the addition of some small low power lasers, and other devices, this station could also assess the scattering properties of the atmosphere in the vicinity of the laser, and thus provide the basis for determining the fraction of the laser's power that would be scattered, and thus the brightness of the laser.

Pre-launch inspection of all satellite payloads could determine the presence of a reactor core. This would require placing a radiation monitoring device next to the exterior of each launcher's payload shroud shortly prior to launch. This would not require actual viewing of the satellite, and thus would not compromise the design characterisitcs of the payload.

And finally, some limits may require creative approaches to verification. It may be difficult to distinguish prohibited anti-missile technologies from benign scientific endeavours. The participation of the international scientific community in the development and execution of projects, such as large space-based astronomical telescopes or nuclear powered planetary probes, could provide reassurance that they were not being used as a cover for military developments.

Recommended Thresholds

Limit 1 - Laser Brightness

Directed energy systems such as lasers should have a limit of 10^{19} watt and joules per steradian on their peak and average brightness, which is a function of the laser's power and energy, as well as the laser's wavelength and the diameter of the primary beam director mirror. Brightness is the most useful measure of a laser's performance. Brightness levels needed for effective ABM systems would probably be hundreds or thousands of times higher than the proposed threshold. Even though the maximum brightness of American military lasers has increased at the rate of a factor of 100 every five years since the early 1970s, the proposed threshold would provide a five to ten year lead-time protection. The minimum brightness level required for ABM purposes is about 10^{16} watt per steradian, and no other applications require lasers of this brightness, except for anti-satellite weapons. However, lasers of such brightness may be relatively small and difficult to identify. Verification of such a limit would probably require the use of non-intrusive in-country monitoring stations located near identified or suspected laser facilities.

Limit 2 - Interceptor Kinematics

A ban on the testing of ABM interceptors (defined as the approach within 10 km at a relative velocity in excess of 10 metres/second) above an altitude of 40 km would preclude the further development exo-atmospheric interceptors for area defence. This would also ban anti-satellite weapons. Systems tested below 40 km with a relative velocity in excess of 4 km/second would be subject to the deployment limits of the Treaty, thus reducing concerns about the strategic implications of antitactical ballistic missile systems, while permitting testing of short-range endo-atmospheric ABM interceptors. The 10 metres/second threshold would permit the rendezvous and docking of manned spacecraft, since such vehicles have very low closing velocities during the final several hundred kilometres of the rendezvous. Such a threshold could be monitored by national technical means.

Limit 3 - Radar Deployment

A limit on the total number of permitted large-phase array radar transmitter faces (perhaps the fifteen that both Parties appear to plan) as well as specification of the distance from the national border (for instance 350 km) that construction is permitted and specification of what constitutes a spacetrack radar, would resolve the Krasnoyarsk and Fylingdales issues. And lowering the Treaty's power/aperture product threshold definition of an ABM radar by a factor of ten from 3 000 000 to 300 000 would lessen concerns about anti-tactical ballistic missiles.

Limit 4 - Mirror and Window Aperture

A limit of 5 square metres (a diameter of about 2.5 metres) on the aperture of ground and space based laser beam director mirrors, space-based sensor satellite mirrors, and the windows on airborne telescope systems, would constrain the ABM potential of all these systems, and could be monitored by national technical means. Although the area of beam director mirrors could vary widely, calculations of the performance of space-based systems typically use areas substantially in excess of 5 square metres. This threshold would usefully constrain space-based lasers, which might pose problems for monitoring a brightness limit. Anti-missile sensor satellites require much larger optical systems than simple early warning satellites. Airborne telescopes need much larger windows on the aircraft than are required for astronomy and intelligence collection.

Limit 5 - Ban on Space Nuclear Power Sources

A ban on the testing in space of nuclear reactors would preclude the use of reactors to power space-based ABM sensors. This could be verified by pre-launch inspection of satellites by radiation monitors, as well as by satellite sensors. This should be a complete ban for at least the next 15-20 years, with subsequent reviews in the fashion of the Non-Proliferation Treaty. In the 21st century, exceptions could be made for scientific spacecraft, verified through international participation in the project.

Limit 6 - Annual Mass Launched into Orbit

Agreement not to place more than 300 tonnes of payload into orbit each year would permit both Parties to conduct current and projected space projects, while providing reassurance that a space-based defence was not being covertly deployed.

Part II

Prevention of Proliferation of Nuclear Weapons

Preventing Proliferation of Nuclear Weapons: Hopes and Realities

Esmat Ezz

Introduction

The dangers of nuclear weapons to mankind and to the survival of human civilization has been recognized almost immediately after their use in Hiroshima and Nagasaki in 1945. The United States had a monopoly on these weapons until the Soviet Union exploded its first atom bomb in 1949, marking the start of a nuclear arms race between the two major powers. The proponents of the development of the H-bomb by the United States believed that if the USSR got there first, it would acquire "world supremacy". So, the USA exploded its H-bomb in 1952 and the USSR followed in 1953. The USA was concerned to match what they feared the USSR might do. The USSR then proceeded to match what the USA had done. Nuclear weapons became a symbol of power. Eugene Rostov, the previous head of the Arms Control and Disarmament Agency, once said "The nuclear weapon is primarily a political, not a military force - a potent political force, generating currents of opinions which are transforming our world"[1]. The United Kingdom in turn exploded its atom bomb in 1952 and its H-bomb in 1960, followed by France which exploded its atom bomb in 1960 and its H-bomb in 1968. China exploded its atom bomb in 1964 and its H-bomb in 1967.

Worried about the consequences of the spread of nuclear weapons and the nuclear arms race, a group of eminent scientists headed by Bertrand Russell and Albert Einstein, issued their Manifesto in 1955, in which they called for scientists to "assemble in conference to appraise the perils that have arisen as a result of the development of weapons of mass destruction". On 7 July 1957, a group of the world's most distinguished scientists - 22 individuals from 10 countries - met in the Canadian village of Pugwash, Nova Scotia, in response to the Russell-Einstein Manifesto, to discuss ways to reduce the dangers of nuclear war and find peaceful means for the resolution of international disputes. This marked the start of the Pugwash Movement[2].

Nuclear Weapons and National Prestige

Possession of nuclear weapons has been considered by some politicians and national leaders as evidence of power and national prestige. Some analysts argue that China was recognized as a great power when she went nuclear. India's nuclear power is regarded as a symbol of strength which rates her next to the five nuclear powers. Although there is some truth in this, it should be pointed out that military strength, whether conventional or nuclear, is just one aspect of the overall strength of a country which is basically socio-economic, cultural and moral. When it comes to prestige or overall strength, one cannot claim that nuclear India is stronger or more prestigious than the non-nuclear Japan, or that the non-nuclear Federal Republic of Germany is weaker than nuclear China.

In the 1979 BBC Reith lecture Professor Ali Mazrui argued that the main African States, Nigeria, Zaire and an ultimately black-run South Africa, should move towards possession of nuclear weapons. That, in his opinion, would be a recovery of adulthood for Africa and perhaps an incentive for the present nuclear powers to become more serious about disarmament[3]. But in the same lecture, Mazrui said that Africa's weakness is mainly due to three things: technological underdevelopment, organizational incompetence, and military impotence. He proceeded to say "African countries manufacture very little themselves and are in command of only the most rudimentary technological capability. Africa's organizational incompetence is aggravated by political instability and social corruption". One may ask, if the situation is as he mentioned, is acquisition of nuclear weapons the solution? Is it feasible? At what cost, and for what purpose? The logical and sensible answer to this suggestion was given by Folahinmi Olabode Adisa, "To demand that Nigeria should divert billions of Nairas, as India is doing, leaving 90 per cent of the population in deprivation, is not only unjustified, it could be criminal. But the issue is not just bread and butter versus guns, for a number of reasons. Firstly, there is no plausible security threat to justify such line of action. Besides, improving the lot of the Nigerian population is a security device. Moreover, it is difficult, if not impossible, to erect any military structure on a weak and archaic socio-economic foundation"[4].

Can Nuclear Weapons Ensure Peace?

The advocates of nuclear weapons claim that they ensure peace by deterrence. The French strategist, Pierre Gallois, evolved a thesis that far from trying to stop the spread of nuclear weapons by treaties, it was more sensible to encourage the spread of nuclear weapons, and to let everybody who wanted them have them. He argued logically, that the USA and USSR had not gone to war for over twenty years, largely because their nuclear weapons deterred them from doing so. They were conscious of the threat of mutual assured destruction if nuclear weapons were invoked. Thus, he argued, that if nuclear weapons were injected into every regional conflict, the Middle East, South Asia, white versus black Africa, a measure of nuclear deterrence would develop to ensure peace in those parts of the globe[5]. Hedley Bull argued that following Gallois's thesis to its logical conclusion, was like saying that the best way to keep death off the roads was to put a small pile of nitroglycerine on every car bumper. Everybody would drive infinitely more carefully, but accidents would occur, people being human, and cars breaking down, and the results would be nastier. Gallois ignored the fact that conflicts that had occurred in the Middle East, South Asia and Africa, are not really comparable to the "Cold War" conflict between the USSR and the USA in the fifties and sixties. The regional conflicts are qualitatively different[6].

The USSR and the USA have both deployed their atomic weapons in a heavily protected second-strike posture, with expensive reconnaisance, and command and control systems which no country in the Third World could afford. In a conflict where one country has nuclear weapons while the other side has not, there would be always the temptation to use those weapons. In a conflict where both sides have nuclear weapons vulnerable to the other side's aggressive strike, each side will be tempted to strike first to destroy the other side's potential. J.E. Spence puts the analogy to that of two gunmen in a Western, the gunman who draws first and faster, and shoots more accurately, wins[7].

The Concept of Nuclear Deterrence as a Basis of Security

This concept is being questioned by most analysts, because for the concept to be effective the nuclear powers must in fact be prepared to fight a nuclear war. This concept breeds conflict and political confrontation and entails the risks of fatal misunderstandings and mistakes[8]. I shall try to analyse the situation in two countries which adopted this concept of nuclear deterrence as a basis of their security, namely South Africa and Israel. In the early seventies, George Quester argued that "by manufacturing nuclear weapons itself, South Africa seemingly would stand to gain less than it would loose. Its conventional superiority over any political opponents in Africa is so clear that it is hardly seen advisable to change the rule of the game"[9]. These views were shared by other analysts[10], and apply exactly to Israel which has been maintaining conventional military superiority over all the Arab countries combined. This situation constitutes the declared policy of the United States of America, the unconditional and constant supporter of Israel. Despite this, Israel has adopted the concept of nuclear deterrence as a basis for its security. There is no doubt that both these countries have nuclear weapons. In the case of Israel it has been reported[11] that it is also stockpiling the materials necessary for assembling hydrogen bombs. Discussing nuclear proliferation possibilities of the "Pariah" states, Harkavy mentions that one must consider a range of strategic doctrines; last resort deterrence; deterrence by uncertainty; massive retaliation; "trip wire" doctrines; tactical battlefield use; etc. "Pariah" states must be concerned with the big powers responses, ranging from nuclear pre-emption to withdrawal of economic and political support; encouragement of rival nuclear development; possible usefulness of nuclear deterrence in guerilla warfare; deeper isolation.

Countries which adopt the concept of nuclear deterrence as a basis of their security, and acquired nuclear weapons, should realize that the quest for a significant margin of military superiority is an illusive one. If the objective is to inspire fear in their opponents, it is doubtful whether this would assure their security, but it would certainly enhance hatred, mistrust and invite hostility. Every nation has the inherent right of individual or collective self-defence in face of armed attack, but security for one side cannot be sought at the expense of that of the other side. The military superiority of South Africa and Israel, including their nuclear arsenals, cannot now or in the future guarantee their security or lasting peace. It is impossible to think that nuclear weapons can furnish the required safety and security for the men, women and children in South Africa and Israel who have the right to live happily without fear or threat. The logical route for security is not nuclear deterrence or military superiority, but it is through the recognition of the legitimate rights of the nationals in South Africa and the Palestinians in the Occupied Territories. The so-called "Pariah" states should learn to live and let others live and exercise peaceful coexistence with their neighbours.

Technology Necessary for Nuclear Weapons Production

The technology to construct a nuclear weapon, according to a variety of studies is increasingly widely available. But for a country to become a real nuclear power, it requires - besides the poltical will to go nuclear - advanced technology, and substantial economic resources. Diffusion of the nuclear power technology to many semi-industrial and non-industrial countries is accelerating their rate of acquisition of skills, facilities and materials relevant to nuclear weapon development.

It is generally accepted, that a modest and less sophisticated programme could be sufficient to develop and produce a small number of simple, but militarily useful nuclear explosives. A country could build a simple small reactor fuelled by natural uranium, together with a small reprocessing plant to extract plutonium from the spent fuel. Such facilities were thought to be much less difficult to construct and less costly (at least by a factor of 10) than a commercial nuclear power plant together with even a small reprocessing plant[12].

Materials from commercial nuclear power reactors could be diverted for military purposes, but such materials would normally not be suited for weapon design. Power reactors produce plutonium with significant amounts of plutonium-240, which is troublesome for bomb design, as it raises the critical mass and could fission spontaneously. There is no simple feasible way of separating Pu-240 from Pu-239. Furthermore, the international safeguard arrangements and political commitments would inhibit most countries from diverting materials from power reactors. Consequently, even countries with existing nuclear power programmes would probably build special reactors if they wanted to produce plutonium for weapons.

Enriched uranium is the other fissile material for nuclear weapons. Theoretically, 10 per cent enriched uranium is weapon-grade, but in practice at least 50 per cent enriched is required, and even then the critical mass would be three times that needed in pure uranium-235.

There are three approaches to reduce the likelihood of non-nuclear weapons states drawing on nuclear technology to produce weapons: banning export of the sensitive components of the fuel cycle, reprocessing and enrichment technology; exacting assurances that such technology will not be used for weapon purposes; and reducing the nuclear power related incentives to acquire such technology.

While the USA refuses to permit export of enrichment and reprocessing technology to non-nuclear weapon states, some other countries do not. West Germany has included a uranium enrichment plant and a fuel reprocessing plant in its package deal with Brazil. France was about to conclude the sale of a reprocessing plant to South Korea, but the USA exercised pressure on South Korea to forgo that deal.

Treaties, Conventions and Agreements Controlling Nuclear Weapons

1. The 1959 Antarctic Treaty, provides for the demilitarization of Antarctica. It is the first treaty to put into practice the concept of nuclear-weapon-free zones.

2. The 1963 Treaty Banning Nuclear Weapons Tests in the Atmosphere, in Outer Space and Under Water (Partial Test Ban Treaty). It does not prohibit underground tests.

3. The 1967 Treaty on Principles Governing the Activities of States in the Exploration and Use of Outer Space, including the Moon and Other Celestial Bodies (Outer Space Treaty). It bans nuclear and other weapons of mass destruction from Earth's orbit.

4. The 1967 Treaty for the Prohibition of Nuclear Weapons in Latin America (Treaty of Tlatelolco). A multilateral regional arrangement, creating the first nuclear-weapon-free zone in a densely populated area, it was the first arms limitation agreement to provide for verification by an international organization.

5. The 1968 Treaty on the Non-Proliferation of Nuclear Weapons (NPT). It aims at the prevention of the spread of nuclear weapons to non-nuclear-weapon countries, at promoting the process of nuclear disarmament, and guaranteeing all countries access to nuclear technology for peaceful purposes.

6. The 1971 Treaty on the Prohibition of the Employment of Nuclear Weapons and other Weapons of Mass Destruction on the Sea-Bed and the Ocean Floor and in the Subsoil Thereof (Sea-bed Treaty). It bans the placement of nuclear and other weapons of mass destruction and facilities for such weapons on or under the sea-bed outside a 12-mile coastal zone.

7. The 1974 Treaty on the Limitation of Underground Nuclear-Weapon Tests (Threshold Test-ban Treaty). It establishes a nuclear "threshold" by prohibiting underground nuclear-weapon tests with a yield exceeding 150 kilotons.

8. The 1976 Treaty on Underground Nuclear Explosions for Peaceful Purposes (Peaceful nuclear explosions Treaty). It prohibits the carrying out of any nuclear explosion for peaceful purposes having a yield exceeding 150 kilotons, or any group of explosions with an aggregate yield exceeding 1500 kilotons.

The last two treaties have not entered into force, but each party has agreed to adhere to its substantive provisions as long as the other does likewise.

9. The Nuclear Suppliers Group, which consists of 15 supplier countries including Western, Eastern and neutral countries. It was established in 1975 and prepared rules and controls over nuclear supplies restricting the export of sensitive material and equipment. The group urged that full scope safeguards should apply to all nuclear exports. However, only Australia, Canada and Sweden rigidly enforce full scope safegurads over all their nuclear exports to all non-nuclear countries[13]. The efforts of the Suppliers Group were expanded in 1977 in the International Nuclear Fuel Cycle Evaluation (INFCE). Sixty-six countries studied the measures that could be taken to minimize the danger of nuclear proliferation without jeopardizing peaceful programmes.

10. The INF Treaty between the United States of America and the Soviet Union in which both sides agreed to eliminate a whole category of nuclear weapons - the medium range missiles within three years and the short range missiles within 18 months. It is true that it eliminates only a small part of the nuclear arsenals of both countries, but it was universally welcomed as a sign and hope. A sign that the two major powers are moving in the direction of the elimination of nuclear weapons through negotiation in good faith, and a hope that serious steps be taken in conformity with their obligations under article VI of the NPT.

Has the NPT Prevented Nuclear Weapon Proliferation?

The non-nuclear-weapon states joined the NPT and agreed to give up the nuclear option in exchange for an end to all nuclear testing and an obligation by the nuclear powers "to pursue negotiations in good faith on effective measures relating to cessation of the nuclear arms race at an early date and to nuclear disarmament" (Article VI).

The treaty entered into force in 1970 and now has 140 parties. It is generally regarded as the most important treaty in the field of nuclear arms control and as a pillar of the non-proliferation regime. Many analysts consider that the treaty has been remarkably successful in preventing the horizontal proliferation of nuclear weapons. **But is this true? Actually, it has prevented horizontal proliferation only among its parties.** Most of the non-nuclear-weapon parties to the NPT have neither the technical nor the economic capabilities required to produce nuclear wepaons; a few have the capabilities, but they chose not to produce nuclear weapons as a matter of principle or for other reasons. Countries that have the desire to produce such weapons, and have or will soon have the capability to do so, did not join the NPT. India exploded its nuclear device in 1974. Reports have been accumulating about the size of the nuclear arsenal of Israel. Similarly, the nuclear capacity of South Africa has been documented. Reports about other near-nuclear countries have mentioned Argentina, Brazil, and Pakistan. It is common knowledge, that most if not all these countries have received some direct or indirect help in nuclear technology, know-how, or even critical nuclear materials from nuclear powers party or non-party to the NPT.

Furthermore, the NPT does not prevent or prohibit non-nuclear parties from developing a nuclear weapons capability. It merely prohibits them from acquiring or manufacturing the actual nuclear weapons themselves. Therefore, a party to the treaty could stockpile plutonium or enriched uranium and then withdraw from the treaty after giving three months notice, and produce nuclear weapons.

With respect to vertical proliferation: cessation of the nuclear arms race and ultimate elimination of nuclear weapons, the nuclear powers have not lived up to their obligation under the Treaty. Vertical proliferation and the nuclear arms race have been going on since the NPT entered into force in 1970.

The non-nuclear-weapon states have regarded a comprehensive nuclear test ban as the centre-piece of the non-proliferation regime, the most important, most feasible and most easily attainable single step to halt the further proliferation of nuclear weapons by both the nuclear and the non-nuclear states. They also regard it as a test of seriousness of the intentions of the nuclear powers to halt the nuclear arms race as provided for in the NPT[14]. The Soviet Union, United States and United Kingdom have undertaken to seek to achieve the discontinuance of all testing in both the 1963 partial test ban Treaty and the 1968 Non-Proliferation Treaty. But some nuclear powers are not willing even to start negotiations for a comprehensive test ban treaty despite their previous commitment to do so.

In the Third NPT Review Conference, this issue was, as in the previous Review Conferences, one of the main outstanding problems. The Conference, except for certain states, deeply regretted that a comprehensive test ban treaty has not been concluded so far, and called on nuclear-weapon states party to the NPT to resume trilateral negotiations, and on all nuclear-weapons states to participate in the urgent negotiation and conclusion of such a treaty as a matter of the highest priority in the Conference on Disarmament (CD). In the 1986 CD Session, no consensus had been reached for the third consecutive year. Members of the Group of 21, stressed the urgent need for a comprehensive treaty on the complete prohibition of testing of all types of nuclear weapons in all environments by all states. They believed that continued nuclear-weapon testing intensified the nuclear arms race and that there was no valid reason for delaying the conclusion of such treaty, since the existing means of verification were adequate to ensure compliance with a nuclear test ban. Members of the Group drew attention to the Five Continent Peace Initiative (Argentina, Greece, India, Mexico, Sweden and the United Republic of Tanzania) for putting an end to all nuclear testing, as well as the offer to lend their good offices to establish verification mechanisms to monitor such a moratorium. The socialist countries expressed their readiness to agree to any format of work which would permit the research for the solution of all the problems involved in a comprehensive test-ban treaty. A group of Western countries pointed out that, in their view, the inherent technical problems of the indispensable prerequisite of a future comprehensive test ban - namely, verification and compliance - had not yet been solved in their entirety, and therefore further preparatory work was required. The United States reiterated that a nuclear test ban remained its objective to be achieved in due course, in the context of significant reductions in the existing nuclear arsenals, and repeated its objections to moratoria on nuclear testing. Despite the attitudes of some nuclear powers, a comprehensive test ban is the practical means for the gradual elimination of existing nuclear weapons and prevention of both vertical and horizontal proliferation.

The CD, in its 1988 session, has adopted an agenda which includes the discussion of the subjects of a nuclear test ban, cessation of the nuclear arms race, nuclear disarmament and prevention of war. It is hoped that the atmosphere that led to the conclusion of the INF treaty will prevail and positive results will be achieved.

Security Assurances for Non-Nuclear-Weapon States

In the framework of the NPT there is no guarantee from the nuclear-weapon states not to use or threaten to use nuclear weapons against the non-nuclear weapon states parties to it. This was and still is a major concern for the non-nuclear-weapon states who willingly relinquished their nuclear option. It also raised much doubt about the value of the Treaty. The Security Council resolution 255 (1968) on nuclear guarantees welcomes the intention expressed by certain states (USSR, USA and UK) to provide or support immediate assistance, in accordance with the Charter, to any non-nuclear-weapon state party to NPT that is the victim of an act or an object of a threat of aggression in which nuclear weapons are used. This resolution does not offer a sufficient guarantee to non-nuclear-weapon states. Similarly, the unilateral statements of nuclear powers made at the First Special Session of the General Assembly devoted to disarmament in 1978 does not satisfy the security requirements of the non-nuclear-weapon states.

The Conference on Disarmament, in its 1986 session, held consultations in order to explore ways and means to overcome difficulties encountered in carrying out negotiations in connection with the question of "Effective International Agreements to assure the non-nuclear-weapon States against the use of Threat to use nuclear weapons".

The Group of 21 continued to believe that the most effective guarantee against the use or the threat to use nuclear weapons was nuclear disarmament and the prohibition of nuclear weapons.

A group of socialist countries shared similar views, while the Western delegations, including those of the three nuclear-weapon states (France, UK and USA), drew attention to the unilateral declarations made by those three states, which they regarded as credible and reliable and amounting to firm declarations of policy. In China's opinion all nuclear-weapon states should undertake not to use or threaten to use nuclear weapons against nonnuclear weapon states. It reiterated that it unconditionally assumed such an obligation and supported all efforts conductive to reaching an agreement on an effective international agreement in that regard.

The DC, in its 1988 session, has set up an **ad hoc** committee to deal with the subject of security guarantees for the non-nuclear-weapon states. It is hoped that positive satisfactory measures will be adopted.

Nuclear-Weapon-Free Zones

A declaration on the denuclearization of Africa was adopted by the Organization of African Unity in 1964. The United Nations General Assembly has repeatedly called upon all states to respect the continent of Africa as a nuclear-weapon-free zone and has condemned the nuclear capability of South Africa.

The 1967 Treaty for the Prohibition of Nuclear Weapons in Latin America (The Treaty of Tlatelolco) established a nuclear-weapon-free zone in Latin America. The 25 Latin American countries who are full parties to the Treaty have undertaken not to test and acquire nuclear weapons themselves, or to permit any other powers to do so, or to station or to deploy them on their territories. Under Protocol I, states that control certain territories within the zonal limits of the Treaty - namely France, the Netherlands, UK and USA - would agree to extend nuclear-weapon-free status to those territories. This protocol has been ratified by the Netherlands, UK and USA, while France has declared that it will take an appropriate decision in due course. Under Protocol II, all nuclear-weapon states have undertaken to respect the nuclear-free status of Latin America and not to use or threaten to use nuclear weapons against the parties to the treaty.

Nuclear-weapon-free zones have also been proposed for the Balkans, Central Europe, Northern Europe, the Mediterranean, the Middle East, South Asia and South Pacific. With regard to the Middle East, there has been a consensus support for General Assembly resolutions urging the parties directly concerned to establish a nuclear-weapon-free zone in that region.

The concept of nuclear-weapon-free zones could be a good step towards the elimination of nuclear weapons from the world. It is of particular importance in areas of tension or where there is a conflict of interest between the major powers. Two elements are essential with respect to such zones:

the nuclear-weapon states should respect the non-nuclear weapon status of such zones and not use or threaten to use nuclear weapons against parties of the zone;

no country in the zone should have or intend to have nuclear weapons; all parties should put their nuclear facilities under full safeguards and accept full verification measures.

The question remains, how is it possible to establish a nuclear-weapon-free zone in Africa with nuclear South Africa, or a nuclear-weapon-free zone in the Middle East with a nuclear Israel?

The Role of Africa in Nuclear Proliferation

In this context, we will consider nuclear weapon proliferation in its broad meaning, to include vertical proliferation, horizontal proliferation and deployment of nuclear weapons in non-nuclear weapon states and even the transit of nuclear weapons through non-nuclear countries. Africa has a major role in nuclear proliferation in this broad sense.

Vertical Proliferation. Africa has more than one-third of the reasonably assured uranium resources in the price range of 80 dollars/kg (about 534 500 tonnes out of the non-communist world total of 1 468 000 tonnes. In the price range $80-130/kg, Africa has 149 000 tonnes out of the non-communist world total of 575 000 tonnes). This huge uranium resource would contribute to vertical proliferation if it became available to nuclear powers. If some restrictions were imposed on these resources, some sort of limitation of vertical proliferation could be achieved. The difficulty is that some major nuclear powers could proceed with vertical proliferation without African uranium.

Horizontal Proliferation. African countries which have substantial uranium resources could be tempted to go nuclear if the technical know-how is developed locally or imported from abroad. South Africa is a typical example. Also countries which produce uranium could contribute to horizontal proliferation by supplying unsafeguarded uranium to other countries, not members of NPT, or accepting safeguards of IAEA. The agreements between South Africa and Israel and Taiwan are such examples. South Africa has demonstrated its will to sell uranium and even enriched uranium outside the international safeguards system. This country has made it clear that it intends to market its uranium strictly in accordance with its national interests and that it can bargain over the price and supplies for political as well as economic ends.

Deployment of Nuclear Weapons. The central position of Africa has a huge strategic importance. Most of the world oil passes through the two basic African routes; either through the Suez Canal or around the Cape. Furthermore, Africa controls the southern flank of the NATO countries. The proximity of the Horn of Africa to the oil-rich Gulf States and the southern part of Asia indicates its importance in any future East-West conflict.

Deployment of nuclear weapons in any part of Africa will trigger reactions involving deployment of similar weapons in other parts of the continent. The major military bases in Africa that could be used in a nuclear global confrontation belong to external great powers; USA, USSR, UK and France. Permission to airplanes or ships carrying nuclear weapons to use African air fields or ports, even in transit, would be a contribution to the spread of nuclear weapons around the world, and should not be allowed.

Conclusions

1. Nuclear proliferation has been going on since the USA exploded its atom bomb in 1945.

2. Treaties, conventions and agreements have done little to prevent satisfactorily vertical or horizontal nuclear proliferation.

3. Most of the states party to the NPT, have no real technical or economic potential to produce nuclear weapons. So it is not logical to claim that these countries did not produce nuclear weapons because of the Treaty. Few states party to the Treaty have both the technical and the economic potential to produce nuclear weapons, but they have not done so, either as a matter of principle or because of some other reasons. These countries would have taken the same attitude if they were not parties of the NPT.

4. Countries with real potential and desire to produce nuclear weapons did not join. These countries have been receiving technical nuclear aid and even critical materials from the nuclear powers, parties and non-parties to NPT.

5. The NPT does not prevent or prohibit non-nuclear parties from developing a nuclear weapons capability or option. It merely prohibits acquiring or manufacturing the actual nuclear weapons.

6. The NPT does not provide security guarantees to the non-nuclear-weapon parties against the use or the threat to use nuclear weapons. This problem must be taken care of.

7. The nuclear-weapon parties to the NPT, have not so far honoured their obligations under the Treaty with respect to the conclusion of a comprehensive nuclear test ban treaty and moving towards complete elimination of all nuclear weapons.

8. Nuclear-weapon-free zones could be a good step towards the elimination of nuclear weapons from the world.

9. The ratification of the INF Treaty between USSR and USA was universally welcomed as a sign that the two major powers have started to move towards the possible elimination of nuclear weapons through serious negotiations in good faith.

References

1. F. Blackaby, "Nuclear Weapon Systems: Possible Future Developments", 37th Pugwash Conference, Gmunden, 1987 p. 147.
2. J. Rotblat, "Scientists in Quest for Peace", MIT Press, 1972.
3. Ali Mazrui, "Africa's Nuclear Future", Survival, Vol.XXII, No. 2, March/ April 1980.
4. Folajinmi Adi, "The Nuclear Rational in Nigeria" Traveaux et Documents No. 3, Centre D'Etude D'Afrique Noire, Domaine Universitait.
5. J.E. Spence, "International Problems of Nuclear Proliferation and South African Position", The South African Institute of International Affairs, July 1980.
6. ibid.
7. ibid.
8. Maj Brit Theorin, "Curbing the Nuclear Arms Race", Sub-regional Conference for the World Disarmament Campaign, Jonkopig, Sweden, 1985.
9. G. Quester, "The Policies of Nuclear Proliferation", Baltimore, John Hopkins University Press, 1973.
10. E. Bustin, "South Africa's Foreign Policy, Alternatives and Deterrence Needs", in Nuclear Proliferation and Near-Nuclear Countries, Onkar Marwah and Ann Schuls (eds)., Ballinger, 1975.
11. Time Magazine, 6 October, 1986.
12. T. Greenwood, G. Rathjens and J. Ruina, "Nuclear Power and Weapon Proliferation", Adelphi Papers, No. 130, 1977, The International Institute for Strategic Studies. C. Starr, "Uranium Power and Horizontal Proliferation of Nuclear Weapons", Science, 244 (4651), 1984.
13. W. Epstein, "The Proliferation of Nuclear Weapons and Efforts to Prevent Further Proliferation", 37th Pugwash Conference, Gmunden, 1987, p. 232.
14. W. Epstein, "The Non-Proliferation Treaty is in Jeopardy", 35th Pugwash Conference, Campinas, 1985, p. 139.

Prevention of Nuclear Weapons Proliferation

Valerii Davydov

At a time when there is a real prospect of eliminating the nuclear threat, international nuclear proliferation policy plays a special role. When the leading nuclear powers - the United States and the Soviet Union - reduce their arsenals and are discussing the possibility of their complete liquidation, proliferation from country to country cannot be tolerated.

At the present time there are more than 130 states committed to the Non-Proliferation Treaty (NPT), which involves the obligation not to make such weapons. The great value of the Treaty strengthening international security will continue to rise, bearing in mind the future development of nuclear energy of the world and the growth of knowledge in the field of nuclear physics. By the beginning of the 1990s there would have been more than 20 so-called 'near nuclear countries' capable of making nuclear weapons in the absence of the NPT. And by the year 2000 they would have numbered 40. During the 20 years of its existence the nuclear non-proliferation regime has proved its effectiveness and reliability. No adherent of the Treaty has made its own nuclear weapons. At the same time all these states continue to take part in the international cooperation in the field of the peaceful use of nuclear energy. The NPT serves the interests of both large and small countries, developed and developing ones, nuclear ones as well as non-nuclear. The future growth of the number of adherents, the future enhancement of the universality of the Treaty, and the growth in the reliability of the guarantees provided by the International Atomic Energy Agency (IAEA), ensure the permanence of the NPT.

At the same time, nearly 40 states still stand outside the framework of the Treaty, including two nuclear states, China and France. They include a number of countries having a well-developed nuclear industry - Argentina, Brazil, Israel, India, Pakistan and South Africa. There still exists in the world forces which would like to challenge the international community and acquire nuclear weapons.

According to estimates by Carnegie Fund experts, Israel may already have a potential of 200 nuclear weapons, including thermonuclear ones. South Africa has a potential to produce about 20 nuclear weapons. By 1991 Pakistan might be capable of producing 15, and India could achieve more than 100 nuclear weapons of Hiroshima size. Argentina and Brazil are also capable of developing nuclear weapons at an early date. In addition, a growing interest in the acquisition of nuclear technology and materials, which might be used for military purposes, is being demonstrated by a whole number of parties to the NPT, among them being Iran, South Korea and Taiwan.

At the present time a new dimension of the nuclear non-proliferation problem - a 'missile' one - has become highly visible. All the candidates nearest to membership of the 'nuclear club' spare no effort in the attempt to develop indigeniously, or to acquire, missiles delivery vehicles. In 1987 Israel tested the missile "Jericho-2" with a range of more than 1300 km.

Experts, also believe that South Africa, in cooperation with Israel, is working to produce cruise missiles. India, Pakistan, Argentina and Brazil have far-reaching space programmes which can be reorientated towards military purposes. After Saudi Arabia's acquisition in 1988 of Chinese intermediary range missiles (2000-3000 km), Pakistan may hope to buy similar missiles from China. If previously the development of missiles delivery vehicles was considered to be a very money and time-consuming effort, the spread all around the world of missile technology is now allowing countries to solve this problem easier and faster. Moreover, the availability of ballistic missile technology might provide an incentive for nuclear weapons acquisition.

Is the appearance of some more states having in their arsenals tens of hundreds of nuclear weapons so dangerous, taking into account that the existing nuclear powers already have more than 50 000 nuclear weapons capable of annihilating the whole world? The success of the international non-proliferation policy depends on the answer to this question.

The overwhelming majority of experts, both in the East and in the West, believe that if nuclear weapons were ever used, this would most likely occur in a conflict between states capable of producing nuclear weapons. Taking into account their political and military ties, the United States, and the Soviet Union would probably be drawn into such local nuclear conflict despite their will. Moreover, in case of further proliferation of nuclear weapons, the threat of nuclear terrorism and the anonymous use of nuclear weapons will certainly grow. An unpredictable growth of the nuclear threat would take place, and the possibility of nuclear weapons being eliminated would become unrealistic and unattainable.

The greatest danger of nuclear proliferation leading to nuclear use lies in places of permanent tension in the Middle East, South Africa and in the Southern part of Africa. Since the Second World War the developing world has experienced about 150 conflicts. It is easy to imagine what might have happened if the parties to these conflicts had had in their arsenals nuclear weapons. Experts believe that if Iran and Iraq had been nuclear powers they would have used nuclear weapons, just as they used chemical ones.

Some Western experts still argue that the enlargement of the "nuclear club" would not lead to a growing risk of nuclear conflict. If, say, both parties to a hypothetical conflict have nuclear weapons, some balance between them on the basis of "mutual deterrence" would have appeared. At the same time, they argue, the threat of nuclear attack would have also made impossible the use of conventional forces. In the final analysis nuclear weapons would play the role of a peacekeeper and stabilizer in relations between countries. As proof of this conclusion these experts refer to the mutual nuclear deterrence which allegedly secured the peace between the Soviet Union and the United States.

In reality, nuclear deterrence doctrines do not augment stability and do not work for the security of states. The whole history of Soviet-American relations, especially during the period of the Cold War, testifies that the risk of nuclear conflict was permanent. As an example, one may cite the Cuban missile crisis of 1962 and the intensification of the danger of nuclear war at the end of the 1970s and the beginning of the 1980s.

The risk may arise as a result of actions of contemporary near-nuclear states - in the Middle East, in South Asia and in the Southern part of Africa. In conflicts, involving such states, where conventional forces are being used, nuclear weapons might be used in a surprise attack if one of the parties proved able to develop them while hostilities still lasted. Interference in such local conflicts by the present nuclear powers might entail the danger of global nuclear catastrophe.

The concept of a "peacekeeping" role for nuclear weapons enforced by Western leaders might produce irretrievable harm to the non-proliferation cause. The spread of nuclear weapons throughout the planet would also make extremely difficult the solution of the problem of reduction and liquidation of existing nuclear arsenals. Already at the present stage, the existence of nuclear forces in France, Great Britain and China poses a whole range of problems in the disarmament field. If several dozens of states acquired nuclear weapons the whole idea of the total liquidation of nuclear weapons would have become an utopia.

Critics of the non-proliferation policy sometimes say that the NPT makes permanent the division of states into nuclear and non-nuclear, and in this way violates the principle of equality and justice in international relations. The attempts to prevent further nuclear weapons proliferation, they argue, have at their root not concern about international security but rather the desire to safeguard the **status quo** in the military field and to secure political advantages. Further proliferation of nuclear weapons would, allegedly, provide all states with equal status and this, in perspective, would produce a favourable impact on the evolution of the international relations and on the process of nuclear disarmament. Such abstract speculations on "justice" and "equality" ignore the main question, namely whether this "just" and "equal" world would be able to survive in a situation of global proliferation of nuclear weapons.

Another mistaken view on the problem is worth noting, namely that limited nuclear proliferation would not result in serious negative consequences for international security. Such statements are based on formal logic - if the international security system was able to adapt itself to the existence of five nuclear powers, why can it not adapt to a larger "nuclear club"? But the axiom of the new thinking in the nuclear age is that true security and nuclear weapons cannot coexist, that in the post-war period life on earth was preserved not by the existence of nuclear weapons, but despite their existence. And it is on the basis of national security considerations that the overwhelming majority of states signed the NPT. By acting in this way they underlined that only the global policy of banning nuclear weapons can secure the preservation of life on our planet.

In a situation of growing technological capabilities of states in the nuclear field the main task of international non-proliferation policy is to deal with the problem of how to reduce or eliminate totally the stimuli to nuclear weapon acquisition. This policy of "stimuli dumping" can be effective only in a situation where the world community is uninterruptedly and resolutely moving towards the total elimination of the nuclear factor from the international relations and towards the creation of bases for a nuclear-free world.

The final success of the non-proliferation policy depends to a great extent on how deep and irreversible are the reductions in the deadly arsenals of the nuclear powers. At the same time, the degree of nuclear disarmament depends on the prevention of nuclear proliferation. The balance between the nuclear powers' obligations to disarm and the non-nuclear powers' obligation not to arm is at the base of the global non-proliferation policy. The key interdependence was stated in all the NPT Review Conferences (1975, 1980, and 1985) and in all the General Assembly Special Sessions on Disarmament (1978, 1982, and 1988).

As long as nuclear weapons remain in existence, it is necessary to take all measures to ensure that they do not pose a threat to the states with non-nuclear status. In this respect the obligations of nuclear weapon states, never to use and never to threaten to use nuclear weapons against states that refuse to produce and acquire nuclear weapons and do not have them on their territories, are very important. Committed to the NPT, non-nuclear states show a quite natural interest in the strengthening of their security. The NPT in fact contains certain international security guarantees for non-member signatories. But what is needed is an unconditional renunciation by all nuclear weapon states of the use of nuclear weapons against states not having them.

Considering demands along these lines by non-nuclear weapon states to be quite justified, at the First Special Session of the General Assembly on Disarmament in 1978 the Soviet Union stated that it would never use nuclear weapons against states which refuse to produce and acquire such weapons and which do not have them on their territories.

At the 33rd UN General Assembly the Soviet Union proposed an international convention on the strengthening of security guarantees of non-nuclear weapon states. Despite the support by the majority of states, and despite the repeated calls by the UN General Assembly Sessions to speed up the development of such a convention, the leading Western states still block its acceptance at the Conference on Disarmament in Geneva. At the root of their negative approach lies the desire of militarist circles to have a free hand in nuclear preparations and to reserve the possibility of waging "limited" nuclear wars on foreign territories.

The initiative to strengthen the security of the non-nuclear states must be taken up by those states themselves. The Soviet Union repeatedly underlined that it is ready, even in the absence of similar steps from other nuclear states, to conclude treaties on the non-use of nuclear weapons with any country in the world, without exception. Non-nuclear states are thus able to start a process of putting security guarantees into treaty form by concluding such agreements with the Soviet Union, at the same time addressing other nuclear weapon states with a similar proposal.

Those countries, which undertake obligations not to produce and not to have nuclear weapons on their territories, may quite lawfully demand a treaty on non-use of nuclear weapons against them as a compensation for the refusal to acquire them. If such treaties were concluded, many "near-nuclear" states which keep the "nuclear choice" still open would think twice before taking a decision to produce their own nuclear weapons. The possession of nuclear weapons would weaken their security in comparision with the security of those non-nuclear states which had received guarantees on non-use of nuclear weapons against them.

Nuclear-free zones can play an effective role in the strengthening of the non-proliferation regime. The participation of countries which still refuse to sign the NPT can, to a certain degree, lessen the threat that they will acquire nuclear weapons. In this case nuclear-free zones can become an important additional obstacle in the way of the further spread of nuclear weapons. In the contemporary situation, when the process of the development of various systems of tactical nuclear weapons for use on battlefield continues, non-nuclear weapon countries may make their visible contribution to the curbing of the nuclear arms race by forbidding the use of their territories for nuclear preparations. Article VII of the NPT confirms the right of states to create nuclear-free zones.

Each year, the UN General Assembly Sessions adopt resolutions supporting the creation of nuclear-free zones in Africa, in South Asia, and in the Middle East. International practice already knows the precedents of international legal formulation of nuclear-free zones - the signing by Latin American countries of the Treaty of Tlatelolco in 1967, and the signing by the members of the South Pacific Forum of the Rarotonga Treaty in 1985. An expanding movement for the creation of nuclear-free zones exists now in many regions of the world.

The Soviet Union wholeheartedly supports the efforts of non-nuclear weapon countries aimed at the creation of nuclear-free zones. The Soviet Union joined the Protocol II of the Treaty of Tlatelolco and Protocol 2 of the Rarotonga Treaty. It is well known that the Soviet Union supports this idea also in respect of other regions. The obligations to abide by the Treaty of Tlatelolco were also taken by other nuclear countries - the United States, Great Britain, France and China. But the United States does not recognize the right of the partners in military alliances to be involved in a nuclear-free zone arrangement. The United States, Great Britain and France refused to join the Rarotonga Treaty, although it received the support of the Soviet Union and China. They also oppose the idea of nuclear-free zones in South East Asia, in the Korean peninsula, and show no interest in the creation of nuclear free zones in Africa and in the Middle East.

Unlike the Soviet Union and China, which voluntarily undertook the obligation of no-first-use of nuclear weapons, the United States, Great Britain and France resolutely refuse to act in the same manner. Forced to react to international public opinion, the United States made the official statements that nuclear war cannot be waged and that there will be no winners in it. But all the Pentagon's Operational plans are based on the readiness to use nuclear weapons, especially from foreign territories, where it has its military facilities. The United States, Great Britain and France ignored the proposal of the Soviet Union and India, contained in the Delhi Declaration on the Principles of a Non-Violent and Nuclear-Weapon-Free World, to conclude urgently an international convention banning the use and the threat of use of nuclear weapons until they are liquidated.

Non-nuclear weapon states, both adherents and non-adherents to the NPT, repeatedly pointed to the fact that the non-proliferation regime would be strengthened only if nuclear weapon powers fulfil their obligations to disarm, according to Article VI of the NPT. The overwhelming majority of world states wholeheartedly support the programme of the total liquidation of nuclear weapons by the beginning of the next century - a proposal which was put forward by the Soviet Union. Welcoming the Soviet-American Treaty on INF and the prospects for 50 per cent strategic forces reductions, they are looking for effective obstacles to a potential nuclear arms race. According to their opinion, the possible agreements in the field must be aimed not at the rationalization of the arms race in order to save money and limit its quantitative dimensions, but rather to put a final and decisive end to it, whether in the field of INF missiles, strategic offensive weapons or tactical nuclear weapons.

One of the main ways of moving towards curbing the arms race is a comprehensive test ban. In this case a reliable obstacle would have been created on the road to further development of more and more sophisticated and deadly nuclear weapons. At the same time the conclusion of a Comprehensive Test Ban Treaty would make tremendously more difficult the development of nuclear weapons by the states which would like to acquire them in violation of the non-proliferation regime. The majority of the world nations, greeting the respective Soviet-American talks started in 1987 in Geneva, call for a comprehensive test ban.

Much attention has also been paid by the world community to a ban on production of fission materials for military purposes. In this respect the world nations call for a moratorium on the part of the nuclear powers on plutonium and enriched uranium production for nuclear warheads and placing of production facilities under IAEA control. Such a step on the part of the nuclear weapon powers could create favourable conditions for putting under international control the facilities of the "near-nuclear" countries - of Argentina, Brazil, India, Israel, Pakistan and South Africa.

At the present stage, when the broad absorption of peaceful nuclear technology and the diffusion of necessary technological "know-how" might facilitate a whole range of countries to shift to military nuclear programmes, supporters of the international non-proliferation policy face new tasks - they need not a confrontation, but a detente, not a nuclear arms race, but real nuclear disarmament. Its final success will depend on the development of a global negative approach to the very fact of nuclear weapons possession. The development of such an approach must be stimulated by practical steps by both nuclear and non-nuclear states, both adherents and non-adherents to the NPT. The adoption of further effective measures for the prevention of the proliferation of nuclear threats is one of the most important dimensions in the struggle for the development of comprehensive international security and of a nuclear-free world.

Revision of the Non-Proliferation Treaty

Krishnaswami Subrahmanyam

Forestalling 1995

According to Article X of the Non-Proliferation Treaty, a conference is to be convened in 1995 to decide whether the Treaty shall continue in force indefinitely or shall be extended for an additional fixed period or periods.

As of August 1988, the Treaty has 139 adherents, including a few principalities and states who are non-members of the UN. Among the important non-adherents to the Treaty are two nuclear-weapon powers, China and France; one nation which has conducted a nuclear test but not built up an arsenal, India; one nation which is widely believed to possess a sophisticated nuclear arsenal, Israel; one nation which is believed to be on the verge of transition to nuclear-weapon status, Pakistan; two Latin American nations with fissile material production capabilities, Argentina and Brazil; one nation which is believed to be engaged in clandestine nuclear weapon efforts, South Africa; and other nations, such as Algeria, Burma, Chile, Mozambique, Niger, Tanzania, Zambia and Zimbabwe. Though the list of non-adherents is relatively small compared with the number of states who have acceded to the Treaty, it is worth noting that it includes eight states which already have or are capable of making nuclear weapons. Among those who have acceded to the Treaty are three nuclear-weapon powers and another half a dozen to a dozen nations capable of making the weapons if they decide to do so. Most of the latter category have not presumably done so because they have the protection of a powerful nuclear-weapon state; have nuclear weapons stationed on their soil; train their troops in the use of nuclear weapons; and are likely to get nuclear weapons released to them when immediate hostilities are anticipated. With the new arms control negotiations now being conducted among the two foremost nuclear-weapon powers, it is difficult to anticipate whether there will be any further changes in the attitudes of some of these non-adherent nations, especially when the emphasis shifts to smaller, extremely accurate nuclear weapons which, it is believed, will produce very little collateral damage.

In any case, a noteworthy fact is that the nations which have abstained from the Non-Proliferation Treaty contain nearly 45 per cent of the world's population. Except for perhaps a dozen industrial nations, the others who have acceded to the Non-Proliferation Treaty have no capabilities to produce the weapons. Viewed in a realistic perspective one has to take into account eight significant non-signatories as against perhaps 12-15 nuclear-weapon states and nuclear-capable signatories. According to Arkin and Fieldhouse[1], though the declared nuclear-weapon nations are only five, nuclear-weapons storage, command, control, communication and intelligence facilities which constitute nuclear war fighting infra-structure, have been spread over 62 nations, principalities and territories. All oceans of the world, hopefully with the exception of the Antarctic, have naval vessels with nuclear weapons deployed on them.

Present State of Arms Reduction Efforts

Now that the Intermediate Nuclear Forces Treaty (INF) has been ratified and a start made to eliminate this class of weapons, one hopes that the Strategic Arms Reduction Treaty (START) will be successfully negotiated and a 50 per cent reduction in strategic weapons will also be achieved. Even then the world will have about 15 000 strategic nuclear weapons (6000 each in the USA and USSR and the rest in the UK, France and China) and perhaps another 8000 shorter range tactical nuclear weapons not covered by these treaties. (The Montebello ceiling is 3500 in Western Europe, and one has to assume a similar arsenal for USSR, and some weapons in France, China and Israel). No doubt if these developments are to come about, a claim will be advanced at the 1990 Review Conference on behalf of the depositary powers that they have carried out a significant portion of their obligations under Article VI of the Treaty "to pursue negotiations in good faith on effective measures relating to cessation of the nuclear arms race at an early date and to nuclear disarmament, and on a treaty on general and complete disarmament, under strict and effective international control". At the last three Review Conferences most of the criticism against the depositary powers was directed against their non-fulfilment of this obligation; this was considered by most people as the crucial failure of the NPT. Even in 1990 the critics are likely to point out that the level of total arsenals is higher than it was when the Treaty entered into force in 1970, and though the reductions are to be welcomed they do not constitute a substantial fulfilment of the obligations then undertaken by the depositary powers. It is also not clear whether even when the USA and USSR agree on reduction of their strategic arsenals to 6000 on each side (50% cuts), the other nuclear-weapon powers UK, France and China will join in any further effort at reductions.

On the other hand, there is considerable talk about modernization of current nuclear arsenals both at the strategic levels (Peacekeeper, Midgetman, Trident D-5, air- and sea-launched cruise missiles, and their analogous systems on the Soviet side) and tactical levels, extending the ranges and accuracy of nuclear weapons to gain the capability to implement the doctrine of discriminate deterrence. There is also the possible danger of extension of weaponization into outer space. In these circumstances it may be overly optimistic to anticipate that the 1990 Review Conference will be less strident in its criticism about the depositary powers than the past Review Conferences.

Proliferation to New States

In his Senate statement on amending the Appropriations Bill, on 11 December 1987, Senator Glenn said categorically that very few Intelligence Agencies of the world today believed that Israel did not have a nuclear arsenal. The trial of Mordechai Vanunu in Israel for his disclosues to the Sunday Times provides further corroboration of this belief. In the case of Pakistan, all that the US President could do was to certify that at the moment it did not possess a nuclear explosive, even while recording a finding that the Pakistan Government was involved in attempts at clandestine export of maraging steel and beryllium from the USA. At the same time the USA has agreed to waive the Symington Amendment for Pakistan for 8.5 years; surely Pakistani efforts over these years will not be totally futile. Dr. A.Q. Khan did admit to an Indian journalist, in the presence of a reputed Pakistani editor, in January 1987, that Pakistan did have the bomb. According to Channel 4 TV programme in the UK on 30 October 1987, Argentina has already assembled two nuclear weapons. According to an Israeli author, Benjamin Beit Hallahami, in his book "Israeli Connection" quoted in the Guardian of 14 January 1988, the flash observed by the VELA satellite in

September 1979 was a joint test by South Africans and Israelis of a nuclear shell fired from a 155 mm gun. While the capabilities of Argentina and South Africa are still matters of doubt and dispute, one cannot dismiss the perceptions of various other nations in regard to these developments and consequent impact on their nuclear policies. India's capabilities to assemble an arsenal is not disputed by the Indians who feel that their restraint has so far gone totally unappreciated.

There is also considerable mention in the literature of clandestine trafficking in nuclear fissile materials. Israel diverted fissile materials from the NUMEC Plant, Apollo in Pennsylvania, and hijacked the ship Sheerberg with its uranium cargo. Pakistan has been able to obtain all equipment necessary for its centrifuge process. An attempt to export a second centrifuge plant by a German firm is reportedly under investigation[2]. Currently the German firm NUKEM is under investigation for exporting to Pakistan nuclear waste materials containing plutonium[3], and there are reports of alleged involvement of Swedish and Belgian firms. The Channel-4 TV programme of 30 October cited Admiral Stansfield Turner, the former director of CIA, that there was a black market in enriched uranium and weapons-grade plutonium based on Khartoum in Sudan, and that other countries (including signatories to the NPT, Iran and Iraq) have been engaged in acquisition of weapon-grade fissile materials. The origin of the material was said to be the European reactors under EURATOM safeguards. The Sunday Times of 7 February 1988 carries a report of the transfer of weapon-grade enriched uranium from West Germany to Argentina in 1985. While the authenticity of all these reports is not beyond doubt, here again perceptions of nations and their impact on popular opinion in nuclear-capable non-nuclear states are of great relevance in regard to future developments.

The presence of very small but distinctly traceable amounts of Pu-239 in the New York water supply[4] would tend to indicate a high probability of the plutonium having originated from one of the US military facilities. It has been reported that thousands of pounds of plutonium were unaccounted for at 15 commercial plants regulated by the US Atomic Energy Commission[5]. In one of the more recent accounting, the US Energy Department discovered a loss of 180 kg of enriched uranium from two of its gaseous diffusion plants at Oak Ridge and Kentucky during a period of five months in 1983-84. These are again non-official reports which have not been officially denied.

The Review Conference of 1990

In these circumstances the Review Conference of 1990 may raise some of these issues while considering whether Articles I and II of the Treaty had been faithfully observed by those nations which acceded to the Treaty.

Article III(2) of the Treaty prohibits the provision of: "(a) source or special fissionable material, or (b) equipment or material especially designed or prepared for the processing, use or production of special fissionable material to any non-nuclear weapon state for peaceful purposes" without subjecting them to safeguards. Article I prohibits transfer of "nuclear weapons or other nuclear explosive devices" by nuclear-weapon states to non-weapon states, while Article II prohibits non-weapon states to receive weapons or other explosive devices. The Canadian Government has recently decided to build nuclear-propelled submarines and the US Government has been reported to be agreeable to transfer the technology to build such submarines. Britain is also reported to be making a bid for the contract. Since the reactors on the nuclear-propelled submarines are not for peaceful purposes, such propulsion reactors do not come under safeguards of Article III(2) if the technology transfer to build them is to take place.

Presumably under this provision the Soviet Union has transferred a Victor class nuclear-propelled, but not nuclear-weapon armed, submarine to India. Apart from Canada, Argentina and Brazil have also declared their intent to acquire nuclear-propelled submarines. The impact of the proliferation of nuclear-propelled submarines on international naval strategic doctrines is yet to be analysed. Certainly this issue is likely to attract the attention of the Review Conference of 1990.

Article IV has lost a major part of its significance in the light of the Three Mile Island and Chernobyl accidents. While countries like Japan, France, UK, USSR, China, India, Brazil, Argentina and Pakistan are keen on expanding their nuclear energy generation, there is a noticeable trend away from nuclear energy in a number of other countries, particularly the USA. Nations do not have now as much enthusiasm to acquire nuclear power stations as they did when the Treaty was drafted. The offer of technology under Article IV was considered a bargain in favour of non-nuclear weapon nations at that time. Since the enthusiasm for nuclear power is waning, it is difficult to foresee the same fervour for nuclear power technology in the Review document of 1990, and the charge that there has not been adequate transfer of that technology. Unlike the seventies, there are today very few illusions about civil engineering applications of peaceful nuclear explosions, though there have been reports of the USSR continuing with such peaceful nuclear tests. It is, however, noteworthy that the Treaty limiting peaceful nuclear explosions concluded between the USA and USSR as far back as 1974 is yet to be ratified by the two countries.

The importance of the 1990 Review Conference is not likely to be in the contents of the document it may or may not (as happened at the second conference) come up with, but in its being a prelude to the 1995 Conference which is to decide whether the Treaty is to be continued indefinitely or for a particular limited period. It is obvious that none of the eight important non-signatories has today any incentive to accede to the Treaty. There is no way of divesting Israel of its nuclear weapons, and the USA has signalled that Pakistani nuclear proliferation is a price that the former is prepared to pay for the latter's strategic cooperation in South West Asia and the Gulf. India is unlikely to accede so long as China has nuclear weapons, and its resolve is further strengthened by US permissiveness towards Pakistan and Israel. South Africa presumably feels that the only way it could sustain apartheid against its own majority black population and the surrounding black states is to follow the Israeli example. Argentina and Brazil have refused not only to accede to the NPT but have refrained from bringing into force the Tlatelolco Treaty signed two decades ago. The ability of UK to bottle up with its nuclear submarines the entire Argentine Navy to its ports during the South Atlantic War has left a deep impression on both Argentina and Brazil which are today embarked upon their own nuclear-propelled submarine R&D programmes.

The Tlatelolco nuclear-weapon-free zone has not proved successful, with 50 per cent of the area of Latin America (Argentina, Brazil, Chile and Cuba) not being covered by it. The disclosure that the USA has contingent plans to bring nuclear depth charges to Roosevelt Road in Puerto Rico[6] during hostilities, robs the concept of nuclear-weapon-free zone of much of its significance. The refusal of Western Powers to recognize the Rarotonga nuclear-weapon-free zone and the US discouragement of a South East Asian nuclear-weapon-free zone (which would necessitate the USA vacating its bases in Subic Bay and Clark Air Field) have not enhanced credibility in the concept of nuclear-weapon-free zones. The references in the recent US Report on discriminate deterrence[7] to the need for very low yield highly accurate nuclear weapons for possible use in future wars, is likely to be yet another significant factor influencing the views of developing nations on nuclear-weapon-free zones. The Palme Commission proposal of a nuclear-free Central European Corridor, the proposals for Nordic, Arctic and Balkan nuclear-free zones have not found favour with the West. It would appear as though the concept of nuclear-weapon-free zones is generally losing ground.

One could theoretically consider the possibility of the Nuclear Non-Proliferation Treaty being continued either indefinitely or for a period of time by the present signatories whose number may well exceed 140 by 1995. But by implication it would amount to virtual confirmation of the status of the eight powers remaining outside the Treaty as nuclear-weapon powers. It would then be a world with eleven nuclear-weapon powers. That may not be an intolerable situation by itself, as has been argued by strategists like Pierre Gallois or Kenneth Waltz. But if that were to happen, will Japan, Taiwan, South Korea, Germany and Italy reconcile themselves to a situation less privileged in international order than Israel, Pakistan, or South Africa? What would be the impact on Iran, Indonesia, and Egypt in the Islamic World, and will they tolerate the privileged position of Pakistan?

New Perceptions

During the last two decades there have been significant advances in understanding the problems involved in fighting a nuclear war, its likely climatic consequences and long-term ecological impact of destruction of nuclear power stations. There have been very serious debates on the ethics of nuclear deterrence. While one school holds the view that possession of nuclear weapons by itself is not unethical - only the use and threat of use would be - there is an alternative school that argues that even possession is unethical. It has been suggested that credibility in nuclear deterrence is becoming eroded with improved understanding that once the first few shots were exchanged it might not be possible to exercise effective command and control over nuclear weapons exchange and prevent rapid escalation. There is also growing fear of nuclear weapons and nuclear materials falling into the hands of terrorists. It has been urged that nuclear threats and use of nuclear weapons constitute the ultimate in terrorism. The leaders of both USA and USSR agree that a nuclear war cannot be won and should not be fought. Recently, a view has been expressed that the threat of using weapons that cannot be effectively used to win a war cannot sustain much credibility. Many would dispute this view.

In 1986, 132 nations have voted in the UN that the use and threat of use of nuclear weapons ought to be banned; the USSR and China were among those voting in favour. The Soviet Union has been urging that nuclear weapons ought to be eliminated and the international community should work steadily towards a non-violent and non-nuclear world (Rajiv-Gorbachev declaration of November 1986). President Reagan is in favour of making the nuclear weapons "obsolete and impotent" and liberating mankind from future terror, but he feels that this could be done through weaponization of outer space. Be that as it may, there is a growing world-wide trend against the doctrine of nuclear deterrence, though it is still the operative policy of the major alliance systems. Those who argue that nuclear weapons cannot be disinvented sound like people of earlier generations who felt that smallpox cannot be eliminated. The inexorable extension of the logic that nuclear weapons cannot be disinvented is that the spread of knowledge of making nuclear weapons cannot be curbed either.

There is also growing understanding that war in the industrial world is no longer a viable instrument of policy. Any nuclear power plant hit will result in a Chernobyl. Any chemical plant hit will produce a large-sized Bhopal. So much building materials, clothes, furnishings and materials of daily use in the industrial world are today made of synthetic materials which produce enormous quantities of pyrotoxins when burnt, killing people and damaging extensively the environment. Fire accidents and airplane crashes have proved that the pyrotoxins produced in the burning of synthetic materials - rather than the fire - produce the overwhelming proportion of casualties.

The political consciousness of people all over the world has reached levels such that it is no longer possible to occupy a country and people except at enormous costs far outweighing any likely benefits of the occupation. That is the lesson of all colonial wars, Vietnam, Afghanistan, Kampuchea, Lebanon, West Bank and Zimbabwe. Perhaps this is not yet wholly applicable to the developing world, but it is clearly recognized in the developed world. Consequently, it is likely that an existentialist state of deterrence against initiation of war in the industrialized world will come about even if nuclear weapons were gradually reduced and ultimately eliminated.

An example in this respect is the Geneva Protocol of 1925 on chemical weapons, asphyxiating gases and toxins. No doubt, in situations of asymmetry, when one side could use it without fear of retaliation by the other side, they have been used, e.g. by Italians in Ethiopia, Japanese in China, Egyptians in Yemen, and Iraqis against Iran recently. Yet, as long as there was recognition that there could be retaliation, deterrence prevailed.

Resolution I(1) of the UN General Assembly called for the elimination of four categories of weapons of mass destruction: biological, chemical, radiological and nuclear. Biological weapons have been banned through a UN Convention. Elimination of chemical weapons which were already outlawed by the General Protocol of 1925 is under active discussion in the Committee on Disarmament. A draft for banning radiological weapons has also been discussed in Geneva. Only in the case of nuclear weapons, while 132 nations are in favour of banning them, 17 nations (the NATO nations and their allies) are for continued legitimacy of nuclear weapons in the hands of a privileged few. Weapons which are treated as legitimate instruments of international power cannot be rendered "obsolete and impotent", as President Reagan would like, nor can they be eliminated, as President Gorbachev advocates. The first step in reducing and eliminating nuclear weapons and rendering them "obsolete and impotent" is to start delegitimizing them. The Geneva Protocol which banned the use and threat of use of asphyxiating gases is a model. The main shortcoming of the NPT is that it legitimized nuclear weapons in the hands of a privileged few and licensed unlimited proliferation by them. It vested a mystique in the weapon, made it a potent instrument of international power, and authorized the possessors of weapons to police those who did not produce the weapons. It is not just an unequal treaty; it is inequitous and non-viable as an instrument to stop proliferation. The Preamble does not have a word to condemn the nuclear weapon as an unacceptable weapon of war.

Revision of NPT

In the light of these developments and our experience of the last two decades there is a good case to explore the possibility to revise the Nuclear Non-Proliferation Treaty to make it an effective instrument to carry out deproliferation of the existing proliferated stockpiles. Article VIII provides for amendments to the Treaty. What is needed is a complete revision of the Treaty which would take into account the current realities, attitudinal changes and new thinking on war doctrines.

In 1990 the Antarctic Treaty is to come up for revision. Already a conference of consultative powers is in session in New Zealand to consider the possible and necessary revisions in the Treaty. In a recent book Sir Antony Parsons, the former British permanent representative to the UN, has already voiced the need for the Treaty to be modified. A similar exercise needs to be undertaken in respect of the Non-Proliferation Treaty. This exercise is likely to be more complex than the revision of the Antarctic Treaty. The ABM Treaty is sought to be re-interpreted although it was agreed for indefinite duration. After the Law-of-Seas Treaty was signed by overwhelming majority, some of the drafting nations have sought to reopen some of its clauses for amendment. The Geneva Protocol is to be replaced by a more comprehensive Treaty. Therefore, the revision of a treaty which has not been acceded to by some important nations, and whose effectiveness has not been subscribed to by nations with proclaimed or suspected arsenals, is both logical and necessary.

At present no party to the NPT is prepared to come out with proposals to revise the Treaty since it is feared that it might initiate a process of unravelling of the entire structure of the Treaty. Having started with the presumption that all nations of the world with reasonable scientific and industrial infrastructure would have an incentive to go nuclear, there is considerable satisfaction that the pace of proliferation has been slower; this slow pace is credited by many to the NPT. On the other hand, it is possible that the original assumption might have been wrong. Those who deliberately vested prestige in the nuclear weapons in the Treaty tend to think that nations go nuclear for prestige purposes. The strategy of ambiguity in respect of nuclear capability has demonstrated that Israel, India, Pakistan, Argentina, Brazil and South Africa do not have prestige as the main consideration, though this may have been true in the case of France and possibly of Britain. The argument that proliferation of nuclear weapons (in terms of decision-making centres) would seriously enhance the danger of nuclear war becomes questionable since every nuclear missile submarine constitutes an independent decision-making centre[8]. There are serious doubts whether Articles I, II and VI have not been breached in practice. A nuclear-propelled submarine can sail through Article III. Article IV has lost its attraction. The black market in fissile materials - if at least some of the reports are true - would tend to show that the IAEA safeguards cannot be effective so long as all reactors in the world are not under universally non-discriminatory and foolproof safeguards, and military production reactors may be a source of fissile materials illicitly transferred.

A strategy has to be devised to move gradually from the present NPT (with whatever ambiguous benefits it may have yielded) to a regime which will progressively make the nuclear weapons "impotent and obsolete", reduce the stockpiles gradually, allow the doctrine of nuclear deterrence to decay over a period of time, and finally lead to a non-violent and non-nuclear world. The appropriate time to revise the Treaty is 1995, but the thinking on the revision of the Treaty will have to be initiated in 1990. In fact, the 1990 Review Conference could become significant if it could formulate the broad guidelines on which the revision of the Treaty for 1995 has to be negotiated. It cannot be done in the traditional way of the two major powers, USA and USSR, negotiating a draft between themselves and present a **fait accompli** to the rest of the world. One would need to follow procedures somewhat akin to those adopted in the Law-of-the-Seas Conference, or if it becomes successful the Conference on Antarctica.

The Economic Factor

Besides the above, there is the major issue of the impact of economic factors on the future strategic interaction between major powers. There is a strong section of opinion in the USA which feels that the US defence expenditure must be reduced if the US economy is not to get into a recession. In the USA there is a feeling that West Europe and Japan are having a free ride with USA providing security, and that the latter should spend more on their own security. If Western Europe and Japan are to find more money for their own security will they not find it economical to develop their own independent nuclear arsenals? The recent debate in Western Europe revealed a strong desire on their part to have an independent programme on space. Ideas have been advocated that West Europe could have its own independent nuclear deterrent. This attitude is also implicit in the French and British missiles having been left out of the INF Treaty. Given the likely demographic trends in Europe, nations like Germany may find the choice in favour of smaller land force and a nuclear deterrent. As US hegemony declines, one cannot rule out the possibility of the European allies and Japan developing an increasing trend towards independent security policies.

The Soviet Union is currently offering some options to Western Europe and there are fears in USA and some countries of West Europe that some of those options may be acceptable to Germany. The USSR is under tremendous pressure to cut back on its defence expenditure and devote more resources to the modernization of its civilian economy. They are prepared to offer asymmetric cuts on conventional forces and make a third zero-zero option in respect of shorter-range nuclear weapons. The choice will be largely influenced by economic factors and the consequent relationships between USA and its allies, and the degree of confidence the USSR is prepared to display in cutting back its obsolescent but numerically larger conventional forces deployed in Europe. Developments in this area are bound to have a tremendous impact on the nature and scope of NPT revision.

The Need for a Research Project

One of the major problems we face is that there has not been adequate interaction among the concerned nations on the role of nuclear weapons in current international relations, on nuclear strategic doctrines and their feasibility, on the role of war in the industrialized world, and the motivations of nations to go nuclear. There is an enormous overburden of US literature, produced as a result of mostly intra-American interactions which ignore the sensitivities and perceptions of the rest of the world. Perhaps the same charge could be levelled against the literature produced in the USSR, Western Europe and a few developing countries. In international conferences the trend has been to expound the standard lines rather than to engage in a dialogue and interaction. There is a need for a research project in which all the above issues are posed to policy makers and strategists in different countries and their specific reactions obtained and those again posed back to them for a second round of reactions.

The Pugwash Council should initiate such a project in three workshops which would involve non-official representatives from USA, Canada, Brazil, Argentina, UK, France, FRG, Sweden, Italy, GDR, USSR, Israel, Saudi Arabia, Pakistan, India, Japan and China.

In the first workshop, the present status of nuclear capabilities of the nations concerned, the perceptions on nuclear war doctrines and nuclear deterrence, and the likely situation in 1995, should be discussed. The concerned representatives will then have a chance to discuss the issue with the decision-makers of the countries concerned.

In the second workshop, these responses will be discussed and a comprehensive paper encompassing all the national perceptions prepared.

This will be followed by a third workshop, which will produce a book covering comprehensively the ground and focus on the need to initiate steps to convene a UN procedure on the lines of the Law-of-the-Sea Conference to consider how the Non-Proliferation Treaty should be revised by 1995. Such a volume should be ready before 1990 when the fourth Review Conference is to be held.

It is not the intention at this stage to come up with specific recommendations of the contents of the new Non-Proliferation Treaty; this should be left to the UN Conference. The Pugwash Movement is uniquely equipped to play a major role to help in revising the Non-Proliferation Treaty from an instrument licensing unlimited proliferation for a privileged few and legitimizing the nuclear weapons, into a treaty which will move the world towards a nuclear-free status; one in which nuclear weapons will become "impotent and obsolete". If smallpox can be abolished by united international efforts, and if AIDS is to be fought by total international effort, so should the menace of the nuclear weapons cult.

References

1. W. Arkin and R. Feildhouse, "The Nuclear Battlefields".

2. New York Times, 5 May 1987.

3. International Herald Tribune, 16-17 January 1988.

4. Times of India (Delhi), 28 July 1985.

5. New York Times, 30 December 1974.

6. International Herald Tribune, 14 February 1985.

7. "Discriminate Deterrence", Report of the Commission on Integrated Long-Term Strategy, F.C. Ikle and A. Wohlstetter, January 1988.

8. P. Bracken, "The Command and Control over Nuclear Forces", Yale University Press, 1983. R.C. Aldridge, "First Strike", Boston: Pluto Press, 1983.

A Strategy to Stop the Spread of Fissile Material

Harold Feiveson

Introduction

In 1946, the USA confronted the implications of the indissoluble link between the peaceful and the military atom in formulating its proposals for the control of atomic energy. In its "Report on the International Control of Atomic Energy" (commonly referred to as the Acheson-Lilienthal Report), a Board of Consultants to the Secretary of State's Committee on Atomic Energy formulated its central conclusion as follows[1]:

> "We are convinced that if the production of fissionable materials by national governments (or by private organizations under their control) is permitted, systems of inspection cannot by themselves be made (effective). ...
>
> (A) system of inspection superimposed on **an otherwise uncontrolled exploitation of atomic energy by national governments** will not be an adequate safeguard." (emphasis in original)

The full force of this conclusion has been delayed for over forty years for reasons we discuss below, but it is now about to be severely tested by the decisions of several countries to produce and use separated plutonium in their civilian nuclear power programmes.

The Once-Through Cycle

The problem of ensuring that the uranium and plutonium used and produced in civilian nuclear power reactors are not diverted to weapons has been greatly mitigated by two fortunate circumstances. The first is that under the current non-proliferation regime, almost every facility in non-nuclear weapon states has come under IAEA safeguards.

The second circumstance of note is the predominance in the world's nuclear power programmes of once-through fuel cycles, in which weapons-usable fissile material is never isolated. In a once-through fuel cycle, the U-235 in the fresh uranium fuel entering the reactor is much too diluted (or "denatured") by U-238 to be used for nuclear weapons. For LWRs, the reactor-type predominant in the USA and much of the world, the isotopic fraction of U-235 in the uranium is generally less than about 4 per cent; for heavy water reactors, the type used by Canada and a few other countries, natural uranium is used. Weapon-grade uranium is generally considered to have over 93 per cent U-235.

The plutonium produced in the reactor and contained in the spent fuel discharged from the reactor is not isotopically denatured in the same manner as the uranium. But as long as the plutonium remains intermixed with the highly radioactive fission products contained in the spent fuel, it too cannot be used for weapons. In the once-through fuel cycle, the spent fuel has mostly been left unreprocessed in storage pools at the various individual reactor sites.

This unreprocessed spent fuel would be virtually impervious to theft. A typical spent fuel assembly weighs roughly one-half of a metric tonne and contains about 4 kg of plutonium oxide. It is also intensely radioactive. Two months after removal from the reactor, a spent fuel assembly would deliver in a few seconds a lethal dose of gamma radiation to a person standing next to it without shielding. Thus, with modest physical security precautions, the once-through fuel cycle can be very effectively safeguarded against theft and subsequent use in nuclear explosives by criminal groups and terrorists, even if they possessed considerable skills, resources, and determination.

Although the spent fuel could be chemically reprocessed by a country employing remote handling equipment and shielding, this would take time and, given international safeguards, would probably be observed. While the spent fuel remained in storage pools or other repositories, it could, therefore, be effectively safeguarded to provide assurances that it was not being diverted to weapons.

The Risks in Reprocessing

Despite these strengths of the current non-proliferation regime, there exist, unfortunately, also weaknesses, which are growing in significance. The most evident is that a few important countries - including India, Pakistan, Brazil, Argentina, South Africa, and Israel - have not joined the Non-Proliferation Treaty. These are all states which have not tied their security directly to US or Soviet alliance structures, and in which more or less strong coalition of the military, nationalistic elements, and political leaders have opposed a formal closure of the nuclear option. India, Pakistan, Israel, and South Africa already have nuclear facilities free from international safeguards and have almost certainly accumulated some quantities of weapons-usable material. Each of these countries is also widely thought to have developed some nuclear weapons capabilities. All currently unsafeguarded facilities in non-nuclear-weapon states are in these four countries and in Argentina.

A more hidden but equally ominous risk to the current regime are the substantial efforts now underway in Europe and Japan to separate plutonium from civilian reactor fuel and to recycle the plutonium into fresh fuel for the civilian reactors. France already has a commercial reprocessing plant at La Hague with a capacity to separate about three metric tonnes (3000 kg) of plutonium per year. This reprocessing complex is now being expanded to approximately quadruple its capacity. The British are constructing a commercial reprocessing plant, Thorp, at Sellafield on the Irish Sea, with a planned capacity of about 8 metric tonnes of plutonium per year. Smaller commercial reprocessing plants are under construction in Japan and one is planned, but not yet under construction, in West Germany[2].

The Nagasaki bomb contained 6 kg of plutonium. If the European and Japanese nuclear industries continue with their plans to reprocess, by the end of the century they will have produced over 300 000 kg of separated plutonium - more separated plutonium than currently exists in the weapon stockpiles of the five nuclear weapon states[3]. (By the end of 1986, the USA and Soviet Union each had about 100 metric tonnes of plutonium in their weapon stockpiles; the other nuclear weapon states had much smaller amounts.) By the year 2000 or even earlier, over 25 000 kg of separated plutonium will be separated **annually** from LWR fuel alone. Four countries - France, UK, FRG, and Japan - will together be separating most of this plutonium. Much of the plutonium will be separated from domestic fuel. But France and UK also plan to devote a substantial part of their reprocessing capacity to fuel from FRG, Japan, Belgium, Italy, the Netherlands, Spain, and Switzerland.

The motives for reprocessing have been diverse. For France and Great Britain, commercial reprocessing provided an opportunity to exploit expertise developed in their weapons programmes; it promised them leadership in an area of high technology; and, as long as other countries help to subsidize the reprocessing by sending their spent fuel to be reprocessed, reprocessing has appeared economically attractive.

Initially, reprocessing also appeared a way to rationalize radioactive waste disposal. This, combined with the willingness of France and Great Britain to reprocess foreign fuel, offered a politically attractive way for some countries, including Germany and Japan, to postpone deciding what to do with the wastes being generated in their nuclear programmes, a contentious domestic issue.

It was also widely assumed until recently that the separated plutonium recovered during reprocessing would be used to provide start-up fuel for prototype and commercial breeder reactors. However, because of greatly reduced demand for electricity, the high cost of breeder reactors, and larger than expected uranium resources, breeder-reactor programmes worldwide have slowed dramatically. They will be able to absorb only a small portion of the plutonium currently scheduled to be separated this century. Under these schedules there will be a cumulative surplus of separated plutonium beyond that needed for breeders of at least 100 metric tonnes by 1995 and 200 metric tonnes by the year 2000.

Faced with this surplus plutonium, several countries, led by France, Germany, and Japan, are now planning to recycle plutonium in LWRs. In recycling, the plutonium and uranium contained in the reactor spent fuel are recovered after processing. The plutonium is converted into an oxide and then mixed with natural or depleted uranium oxide to form a mixed-oxide (MOX) blend. The MOX blend is then fabricated into fuel elements which are recycled into fresh fuel for the reactor. The uranium recovered in the reprocessing is also recycled; it is re-enriched (or mixed with uranium enriched to higher than usual levels) and fabricated into regular fuel elements.

In practice, the recycle of the uranium and plutonium discharged from LWRs could reduce natural uranium requirements by about 25 per cent, with the recycled uranium and plutonium each responsible for about one half of this savings. Separative work requirements could be reduced by about 20 per cent[4].

The scope of the recycling now contemplated is staggering. The full maturing of the recycling programmes now planned in Europe and Japan would place over 25 metric tonnes of separated plutonium into routine commerce annually by the end of the century[5]. Several hundred shipments (some by sea or air) would be required each year to transport the plutonium from the reprocessing plants to fabrication facilities to reactors.

The plutonium oxide and mixed oxides involved in the recycling possess a chemical form not ideally suited to the production of nuclear weapons. Nevertheless, plutonium oxide obtained directly or extracted from mixed oxide fuels could be used directly in nuclear explosives or converted into a metal in a relatively short period[6].

The sheer scope and complexity of the emerging plutonium trade will raise formidable difficulties for countries simply to keep track of the plutonium. It will, for example, be increasingly impractical for the USA "to follow" plutonium produced in reactors supplied by the USA or using fuel supplied by the USA. At present, the safeguards agreements attached to such reactors generally give the USA the right to establish the conditions under which the plutonium could be separated and to ensure that the plutonium remains always under international safeguards. But, even today, before the projected expansion in plutonium separation and trade, it has become difficult for the USA to follow the manifold paths of plutonium produced in safeguarded facilities.

The growing scope and complexity of plutonium trade will also make it increasingly difficult to keep plutonium, and other sensitive technologies, useful in producing plutonium, out of international black markets in nuclear materials and technologies. Pakistan is known to have assembled uranium enrichment and reprocessing technologies over many years through clandestine purchases of equipment and components from private companies in Western Europe. Recently, Taiwan was reported to have been working in secret on reprocessing technologies to separate plutonium, presumably using components obtained clandestinely from international markets.

There have been other recent intimations of black markets in nuclear material and technologies. For example, it has been recently reported by the Norwegian government that a significant amount of Norwegian heavy water appeared to have been illicitly diverted in 1983. The heavy water was sold to a West German company but flown instead to Switzerland and there shipped to an unknown destination. A similar multinational diversion of nuclear material through a privately-owned company is now being investigated by the West German government.

Even if the plutonium and other nuclear material and technologies involved in the plutonium trade were safeguarded, reprocessing and recycling would provide a strong foundation for national nuclear weapon programmes. Any country engaged in plutonium recycling would have available potentially huge quantities of readily accessible weapons-usable material. Under such circumstances, if a country had produced all the components of nuclear weapons other than the fissile material cores, it could reduce the time between a decision to acquire nuclear weapons and its achievement on a potentially large scale, from years to weeks. It would also be able to push far towards a nuclear-weapons capability without having to make an explicit decision to acquire nuclear weapons and, in this manner, would be able to hide a latent nuclear weapons programme with an ambitious civilian power programme.

Reprocessing and recycling are now concentrated mostly in advanced industrialized countries which actively support the Non-Proliferation Treaty. However, the emergence of a commercial market in mixed oxide fuels, even if initially restricted to Europe and Japan, will almost certainly, over time, lead other countries to recycle plutonium in their commercial reactors or engage in research with plutonium fuel. The emergence of a plutonium market, in which plutonium fuel will be traded freely within the industrialized countries, will make it extremely awkward for suppliers to deny plutonium fuel to developing countries or to deny them the reprocessing and fabrication facilities able to produce plutonium relatively quickly.

The nuclear industries of the world are kidding themselves that, on top of the division of the world between countries which have nuclear weapons and those which do not, there can now be added another division: between industrialized countries which will be recycling civilian nuclear fuel and developing countries which will not be allowed to do so. They are under the delusion that Europe and Japan, along with the Soviet Union and the USA (if they wish), will be able routinely to recycle tens of thousands of kilogrammes of separated plutonium annually, while the rest of the world docilely accepts that such recycle is not for them.

Reprocessing and Recycling are not Economically Viable

Fortunately, the economic forces driving reprocessing and recycling are far from irresistible. This is partly because nuclear power is in the doldrums. In the USA, there have been no new orders for nuclear plants since 1978 and none are expected before the end of the century, if then. The United Kingdom, West Germany, and Sweden also have effectively stopped ordering new nuclear plants. There is currently less than 10 GWe nuclear capacity operating or under construction in developing countries (excluding Korea and Taiwan). The large size of nuclear reactors compared to the grid capacity in most developing countries, the high capital costs of nuclear power, and the large rural populations not easily reached by interconnected grids are all factors which work against nuclear power in developing countries[7].

Reprocessing and recycling will not be economically justified as long as nuclear power continues to play a limited role in the world's energy picture. Although the current LWR operating on a once-through fuel cycle consumes more uranium per kilowatt-hour than fuel cycles using recycled plutonium, it has much lower capital costs than the breeder, and lower non-uranium fuel cycle costs than fuel cycles using mixed-oxide fuels. This is of particular significance because in today's reactors, uranium contributes only 0.12 cents per kWh or about 3.5 per cent to the total generation cost of electricity.

As a result, the price of uranium would have to rise astronomically before plutonium recycle and the breeder fuel cycle become competitive with the present once-through fuel cycle. The price of uranium would have to reach nearly $350 per kg of uranium - 8 times the 1987 price of uranium - before the cost of LWR generated electricity exceeded that of a breeder whose capital costs are 20 per cent greater than that of an improved design light water reactor. Similarly, plutonium recycle in LWRs could not compete with the once-through fuel cycle until the price of uranium increased nearly fivefold from its present value.

In these circumstances, it is no wonder that the development of a plutonium economy based on the breeder reactor no longer has any urgency. Plutonium recycling in LWRs looks no more attractive. Indeed, whereas the breeder reactor at least has the virtue of burning uranium with great efficiency, plutonium recycling in LWRs would reduce uranium requirements, compared to a strategy of sticking with the once-through fuel cycle, by no more than 20 per cent or so. Savings of this magnitude or greater could be achieved at less cost in other ways without reprocessing and recycling.

It is sometimes said in favour of breeders and plutonium recycling that, economics notwithstanding, they would provide a country greater independence from imported uranium. This purchase of independence, however, would be at the expense of a sharply increased dependence on a steady flow of plutonium from a small number of reprocessing plants, possibly located in foreign countries. In addition, uranium resources, unlike oil, are not concentrated in one region of the world; and the largest deposits are in the politically stable USA, Australia, and Canada. Also, uranium, far more than oil, can be economically stockpiled.

Reprocessing should be Abandoned

The economic realities of reprocessing and recycling create an opening for an international policy initiative to get countries to agree to defer spent fuel reprocessing and plutonium recycling indefinitely (at least until such time as these activities become necessary economically) and to institute effective international control of fissile material and facilities producing fissile material.

It is not too late for a policy to defer reprocessing and recycling. Only France now has a commercial reprocessing programme and, as yet, no country is recycling plutonium on a commercial scale. In at least two countries which are in principle commited to reprocessing - UK and FRG - there is strong domestic opposition to reprocessing, so that even the political parties pushing reprocessing may wish to find an excuse to back off.

France currently finds reprocessing profitable and Britain expects to because other countries (primarily Japan) have in effect financed the French and British operations. Were this to change, reprocessing may look unattractive even to France. Such a change is not out of the question. Countries which have sent their fuel to France and to Britain for reprocessing have done so largely because of internal political factors - notably laws which required the nuclear industries of the country to figure out what to do with spent fuel before nuclear power would be allowed to expand further. But by sending the fuel out of the country, the problem was solved only temporarily. The countries involved will soon have to deal with the high level waste and plutonium which are to be returned to them.

There are, in fact, no grounds to pursue reprocessing to rationalize waste disposal. In terms of repository design for final disposal of the spent fuel, spent fuel and high level waste pose broadly similar problems. The fission product contents are essentially identical and the heat outputs per metric tonne of original uranium are similar for the first hundred years[8]. Although spent fuel would contain more plutonium, significant amounts of plutonium from reprocessing would go into the high-level waste because of process losses during the repeated recycle of plutonium before it fissioned. And, a significant fraction of the plutonium would not be fissioned but transmuted by neutron absorption into heavier trans-plutonium isotopes which would end up in the wastes. As a result, the waste from a plutonium recycle fuel cycle would be about as radioactive for about as long as that from a once-through fuel cycle.

Prospects for a New Gas Turbine

In abandoning reprocessing, countries will be recognizing, more explicitly than many have been so far willing to do, that nuclear power has become an energy choice of last resort. Aside from the complex of rational and irrational factors which have made the public and utilities in many parts of the world wary of nuclear power, during the next few decades at least, market forces alone are likely to favour fossil-fuel alternatives to nuclear power for central station electric generation in much of the world.

One reason for this is the emergence of highly energy-efficient and relatively inexpensive advanced aero-derivative gas turbines now under intensive development[9]. These turbines would be able to produce electricity at costs competitive with coal and nuclear plants in units one-tenth the size. These advanced turbines for central station power generation are spin-offs from advanced military-aircraft jet engines developed in the USA. Possibly the most attractive of the new designs is the intercooled steam-injected gas turbine (ISTIG), which is being developed by General Electric.

The General Electric design would have an output of 110 MW(e) and an efficiency of 47 per cent when burning natural gas. Its estimated capital cost is $400 per installed kilowatt. The main significance of ISTIG is that it offers high efficiency and low cost in a much simpler and smaller machine than current generators. The unit could probably be built in less than two years.

The simplicity and small size of the turbines allow their production to be done mostly in the factory, where costs are easier to control than in the field. This is a principal reason for their low capital cost and short construction time. Their compact, modular construction also make them easy to remove from the field and be returned to the factory for repair (much as is done with jet engines today). Their short construction time and high efficiency at low unit sizes give the turbines great flexibility. Utilities would be able to add capacity quickly as needed. The relatively small scale and low capital costs of the turbines also make them attractive candidates for the small grids of developing countries.

The advanced turbines would initially use natural gas of which there is as much as oil in the world[10] (but is currently used at a half the rate of oil). However, it would also be possible for the gas turbines to exploit the much larger coal resource by matching the turbine unit to a Lurgi coal-gasifier. It would also be possible to use biomass as the source of gas for the advanced turbines.

The high cost and unpopularity of nuclear power in combination with the gas turbine alternative strongly suggests that countries, even those with present commitments to nuclear power, will not undertake significant expansions in their nuclear power programmes over the next few decades. A plateau (or, perhaps, a slow rise or decline) in installed nuclear power capacity worldwide after the year 2000 is therefore the most reasonable prospect for nuclear power. In these circumstances, countries should not find it difficult to give up reprocessing and recycling.

The Greenhouse Effect

Nuclear power is not likely to be rescued by global environmental concerns - including, most prominently, concerns of the greenhouse effect. Nuclear power could contribute to a solution of the greenhouse problem. But, for a very long period, it could do so only marginally. This is principally because over 75 per cent of fossil fuel use in the world today is for non-electric uses[12].

If nuclear power grew slowly as now projected, it could remain indefinitely on the once-through fuel cycle. But, in these circumstances, it could not make an appreciable dent in atmospheric carbon concentrations for many decades. Nuclear power would have to grow manyfold before it could significantly affect the greenhouse problem. Such growth, however, would force countries to adopt reprocessing and recycling on a vast scale. A 1 per cent decrease in the annual carbon input due to the substitution of nuclear power for gas-fired generators would require a scale of nuclear recycling of nearly 1.5 million kg per year. This assumes that breeder reactors would replace ISTIG units. A breeder system involves the annual circulation of about 2000 kg of plutonium. A 1 GWe ISTIG generating plant on natural gas produces about 0.68 million tonnes of carbon per year compared to about 1.7 million tonnes of carbon for a central station coal plant.

Conclusions

For the next few decades, nuclear power growth is likely to be slow or non-existent. Although the greenhouse effect will force the world eventually to cut back drastically on its use of fossil fuels, nuclear power appears destined to play only a small role in this cutback. Conceivably, if alternative energy technologies are developed and commercialized and nuclear energy is allowed to rise or fall through its own economic weight, nuclear power may sink out of sight altogether. In any case, the lull (or demise) in nuclear power has clearly destroyed any economic case for reprocessing and recycling.

Because the present momentum towards reprocessing and recycling has no compelling technical or economic rationale, there is now a window of opportunity for a global initiative to defer (or abandon) these activities indefinitely. Of the countries with strong nuclear power programmes, the USA, Sweden, and Canada have already abandoned plans for commercial reprocessing; the Soviet Union is not known to be pursuing reprocessing on a commercial level; UK, FRG, and Japan do have reprocessing plans but face formidable domestic political opposition; and even France, the world's Pied Piper of commercial reprocessing, may be ready to rethink its position following recent warnings from France's state-run utility that it is unwilling to make further commitments to a plutonium breeder reactor until that venture is more economical.

References

1. A Report on the International Control of Atomic Energy. Prepared for the Secretary of State's Committee on Atomic Energy, U.S. Government Printing Office, Washington D.C., 16 March 1946 (the "Acheson-Lilienthal Report").

2. D. Albright and H. Feiveson, "Plutonium Recycling and the Problem of Nuclear Proliferation", Annual Review of Energy, 1988, 13:239-65.

3. F. von Hippel, D. Albright, and B. Levi, "Stopping the Production of Fissile Material for Weapons", Scientific American, September 1985; H. Feiveson, F. von Hippel, and D. Albright, "Breaking the Fuel-Weapons Connection", Bulletin of the Atomic Scientists, March 1986.

4. The Economics of the Nuclear Fuel Cycle, A Report by an Expert Group, OECD/NEA, 1985, pp. 150, 154-55.

5. D. Albright and H. Feiveson, "Why Recycle Plutonium?", Science, 235, 27 March 1987, pp. 1555-56.

6. J. Carson Mark, T. Taylor, E. Eyster, W. Maraman, and J. Wechsler, "Can Terrorists Build Nuclear Weapons?", in Preventing Nuclear Terrorism, P. Laventhal and Y. Alexander (eds.), Lexington, MA.: Lexington Books, 1987.

7. J. Goldemberg, T. Johansson, A. Reddy, and R. Williams, Energy for a Sustainable World, Wiley, 1988.

8. Nuclear Energy Agency, OECD, Nuclear Spent Fuel Management: Experience and Options, 1986.

9. R.H. Williams and E. Larson, "Aeroderivative Turbines for Stationary Power", Report Number 226, May 1988, Center for Energy and Environmental Studies, Princeton University.

10. C.D. Masters, et al, "World Resources of Crude Oil, Natural Gas, Natural Bitumen, and Shale Oil", Paper prepared for the 12th World Petroleum Congress, Houston, Texas, 1987.

11. V. Ramanathan, "Trace Gas Trends and Climate Change", University of Chicago, January 1987.

Part III

Nuclear and Conventional Forces

The Links Between Nuclear and Conventional Forces

Jean Klein

Historical Review

The question of the links between nuclear and conventional forces has been raised in the United Nations disarmament negotiations from the very beginning. After the bombing of Hiroshima and Nagasaki, the United States resolved that the maximum constraint must be put on arms of mass destruction, and that the elimination of atomic weapons should have priority in the efforts of the international community towards arms regulation. In January 1946, the General Assembly agreed the creation of an Atomic Energy Commission (AEC), and in June the United States presented the Baruch plan which ceded the management of all peaceful activities in the nuclear field to an Atomic Development Authority (ADA) with supranational powers. Once a system of control and sanctions were established, further production of atomic weapons would be stopped, existing stocks would be destroyed and all technological information would be communicated to the ADA. While the declared final aim was the total elimination of nuclear weapons, in the meantime the USA would have maintained its monopoly. It was obviously unlikely that the Soviet Union would accept such a situation. To counter the American plan, Andrei Gromyko requested the prohibition and destruction of all nuclear weapons, as a first stage, but this request was incompatible with the security interests of the Western countries. In their view, nuclear weapons compensated to some extent the superiority of the Eastern side in the conventional forces deployed in Europe.

The next step was the creation in February 1947 of a Commission for Conventional Armaments (CCA) in the framework of the United Nations. It was instructed to submit to the Security Council proposals: (a) for general regulation and reduction of armaments and armed forces; (b) for practical and effective safeguards in the disarmament process. Matters which fell within the competence of the Atomic Energy Commission, i.e. those related to nuclear weapons and the peaceful use of atomic energy, were excluded from the terms of reference of the Commission. Between 1947 and 1950, disarmament negotiations were conducted on both tracks, but the prospects of any agreement were dim. The Soviet Union was involved in a military nuclear programme and would not accept the establishment of a supranational Atomic Development Authority that might spoil the success of this venture; the USA and their allies took comfort from the American nuclear monopoly (until 1949), and were reluctant to renounce this trump card during the cold war. In any case, the international atmosphere was not favourable to disarmament, and the separate treatment of nuclear and conventional forces was an additional obstacle to the conclusion of a balanced agreement. The prohibition of nuclear weapons before the correction of the imbalances at the conventional level would have confirmed the Soviet military superiority in Europe; conversely, maintenance of the American nuclear monopoly (and clear superiority until the mid-sixties) and subsequent reduction of conventional forces, meant permanent inferiority for the Soviet Union. The only way to reach an agreement was to tackle simultaneously the nuclear and conventional components of the military balance, and to combine measures of nuclear and conventional disarmament so that equal security would be maintained at every stage of the arms reduction process. A change occurred in this respect at the beginning of the fifties, during the first phase of dētente following the death of Stalin.

The milestones on this road were the establishment of the Disarmament Commission in January 1952, which replaced the two former commissions, AEC and CCA; and the French-British memorandum of June 1954 on general, progressive and balanced disarmament. This document attempted to find a compromise between the American and Soviet views. It was based on three principles: all the disarmament measures should be conceived and implemented in a sequence that would maintain the security of all the parties concerned during the whole process; the transition from one stage to the next would only occur if the verification procedures had been proved efficient; at every stage measures of nuclear and conventional disarmament would be combined in such a way that a military equilibrium would be kept between all the disarming parties. Progressivity, equal security, and strict control were the features of the French-British plan. In the autumn of 1954, this document was accepted by the Soviet Union as a basis for further discussion. Later, proposals for general and complete disarmament (GCD) were discussed in the framework of the subcommittee of the disarmament commission in London. However, it appeared very soon that the destruction of the nuclear arsenals could not be adequately verified and that the total elimination of this component of the military apparatus was out of reach. The United States then drew the conclusion that general and complete disarmament was a distant prospect, and that priority should be given to the adoption of confidence-building measures in order to create an atmosphere propitious to selective disarmament. The "open skies" plan presented by President Eisenhower during the Geneva summit, in July 1955, reflected the new American philosopy.

This stand was not approved by the French, who maintained the idea of general disarmament in the areas where verification was possible. Jules Moch, who at that time headed the French delegation at the disarmament conferences, expressed his view in a concise formula: "no disarmament without control, no control without disarmament, but all the disarmament which can be verified in an adequate way". The Soviet Union had some reservations about on-site inspections and denounced the Western requests on verification as an attempt to legalize spying; it clung to the idea of general and complete disarmament and did not follow the American suggestion to start the process with confidence-building measures. Nevertheless, the trend to a selective approach of arms regulation was irresistible, and from 1956 onwards, the two great powers made proposals leading to partial arms control measures.

The idea of reducing armed forces and conventional weapons independently from nuclear disarmament was raised by the Soviet Union at the subcommittee of London, and there were discussions on the creation of demilitarized zones in Europe and in the Arctic area. The issue of nuclear tests became sensitive after the accident involving a Japanese fishing boat in the Pacific in 1954, and the growing protests of the non-aligned movement against nuclear weapon tests. In 1957 talks on the limitation of nuclear tests started in London but the subcommittee on disarmament was dissolved at the end of the same year without having achieved anything. In 1958 the three nuclear powers - USA, UK and USSR - decided to negotiate a separate agreement on the prohibition of nuclear tests, and after many tribulations they reached an agreement which contributed mainly to reduce atmospheric pollution but not to stop the arms race: the Partial Test Ban Treaty signed in Moscow on 5 August 1963.

This Treaty illustrates the change which occurred at the end of the fifties in disarmament negotiations, a trend confirmed in the conduct of the two great powers after the Cuban missile crisis in 1962. The aim was no longer a drastic reduction of armed forces and armaments at a global level, and the prohibition and elimination of weapons of mass destruction, but the "control of the arms race".

The theory of arms control stemmed from American studies on the management of modern weapons, particularly nuclear weapons, at a time when the two giants became aware of the risks and the uncertainties of the "balance of terror". In spite of the asymmetries of their nuclear arsenals, each side was capable of inflicting intolerable damage on the other side in retaliation. Since the confrontation between the two systems did not allow a significant reduction of armed forces, it was imperative to find ways and means to prevent a major war which would have meant mutual suicide. Arms control was supposed to contribute to the stability of the deterrence relationship, in preventing the breakdown of the nuclear equilibrium by accident, miscalculation or human error. In this respect it was necessary to define a code of conduct and to take measures to facilitate crisis management; the nuclear capacities had to have only a deterrent function, and an invulnerable second strike capability would discard the temptation of preemptive strikes in an acute crisis. Finally, the dissemination of nuclear weapons was believed to increase the risks of war escalation, and one of the objectives of arms control was to contain and, if possible, to prevent their acquisition by third countries.

The theory and practice of arms control in the sixties and the seventies tended to dissociate nuclear weapons from conventional disarmament for which the great powers lost interest after the failure of the negotiations on GCD. In 1960, the ten states participating in the Geneva disarmament conference put the GCD issue on the agenda, but the difficulties to find a combination of measures covering the reduction of nuclear and conventional weapons, without putting at risk the security of the states concerned, and the enduring quarrels on the verification methods to guarantee the implementation of any agreement, turned out to be insurmountable. The French government, which had tested a nuclear device in the Sahara desert in February of the same year, was not hostile in principle to nuclear disarmament. Jules Moch presented in Geneva a proposal on the reduction of the carriers of nuclear weapons, whose removal could be controlled more easily than the elimination of nuclear warheads and fissionable materials. The Soviet Union supported the French view but seized this opportunity to try to divide the Western allies. After the failure of the summit meeting in Paris, in May 1960, it was clear that the time was not ripe for disarmament, and that the two great powers were more interested in arms control than in real disarmament. General de Gaulle drew the conclusion that the Geneva talks which started in the spring of 1962 in the framework of the Eighteen Nations Disarmament Committee (ENDC) would not be devoted to disarmament but only rubber stamp the decisions taken by the Soviet and American co-chairmen. In order not to be inhibited in the conduct of its foreign policy and the development of an independent nuclear deterrent, the French President chose to practice the "empty chair" policy at the ENDC, but his judgment on the new trends of disarmament were confirmed by the evolution of the Geneva talks during the next ten years: GCD was abandoned in the mid-sixties and bilateralism prevailed in the search for nuclear arms control. The results obtained were quite modest, and agreements such as the Non-Proliferation Treaty (1968), the Seabed Treaty (1971), the Threshold Test Ban Treaties (1974 and 1976), etc., had only marginal value as far as the limitation of arms competition was concerned. With the exception of the Convention on Biological Weapons (1972), all the agreements concluded under the aegis of the Geneva committee concern nuclear weapons. There was no link whatsoever with the reduction of conventional arms, and this question was not seriously considered until it was raised in connection with the convening of a Conference on Security and Cooperation in Europe (CSCE).

When the Warsaw Pact states launched their Budapest appeal to convene an all-European Conference on Security (March 1969), the military issues were not the main topics on the agenda. The emphasis was put on cooperation between European countries at various levels; regional disarmament was hardly mentioned. Conversely, the NATO countries belonging to the military integrated command were in favour of negotiations on mutual and balanced forces reductions (MBFR) and had raised the issue in a communiqué adopted in Reykjavik in June 1968. At that time, the Warsaw Pact states were reluctant, and their reservations subsided only when it was clearly stated by the Western side that the opening of talks on force reductions in Europe was one of the conditions to its participation in the CSCE. In June 1970, the Warsaw Pact states agreed to discuss the reduction or withdrawal of foreign troops stationed in Europe, and one year later Leonid Brezhnev, in his Tbilisi toast (14 May 1971), hinted a negotiation encompassing national as well as foreign troops but raised indirectly the question of nuclear weapons. During the Moscow summit, in May 1972, Secretary Brezhnev and President Nixon supported the idea of a CSCE and agreed that "reciprocal reductions of armed forces and armaments first in central Europe would contribute to the stability and security in Europe". The preliminary talks on a CSCE started in November 1972 and a few months later the so-called MBFR process began in Vienna. The negotiations have been going on since October 1973, but the chances to conclude an agreement are limited and there is a consensus to replace the MBFR talks by negotiations on conventional stability after the conclusion of the Vienna follow-up meeting of the CSCE.

At the present stage of negotiations, I will confine myself to some general remarks on the linkage between nuclear arms control and the attempts to limit or reduce conventional weapons, especially in Europe, as this is the only area where the question has been addressed in concrete terms. As the question of nuclear weapons has also been raised in the framework of the MBFR talks, and is discussed in connection with the elaboration of a mandate for conventional stability talks (CST), it might be useful to clarify the positions of the interested states.

Linkage with Nuclear Arms Control

The regulation of conventional armaments has been discussed since the beginning of disarmament negotiations, and as we have seen, the United Nations tackled the issue first in a separate way, then in combination with nuclear disarmament, the frame of reference being the model of general and complete disarmament. Simultaneously, there were attempts to find regional solutions, and in the fifties a great variety of disarmament and/or disengagement plans were proposed by political leaders, retired diplomats (George Kennan) and governments, to diminish the threat of military confrontation on the continent. These plans raised objections in NATO circles and were considered incompatible with the implementation of deterrence strategies. Some experts argued at that time that the withdrawal of tactical nuclear weapons would break the chain of the NATO triad - strategic weapons, tactical nuclear weapons and conventional forces - and undermine the credibility of the American extended deterrence. Another argument was that the removal of nuclear weapons would give prominence to the imbalances between NATO and Warsaw Pact conventional forces. Thus, it was necessary to correct these imbalances before, or at least in parallel with, any limitation of nuclear forces on the Western side.

In 1972 the Soviet Union and the United States reached an agreement on the limitation of their strategic arms (SALT I) which was considered by the Western allies as confirming the strategic parity and weakening the nuclear guarantee offered in the framework of NATO. During the SALT negotiations the European countries objected also to the inclusion of the so-called forward-based system (FBS) in the negotiations on strategic weapons, since these systems were devoted to the security of Europe and could not be disposed of without the consent of the allies. They got satisfaction but the question was raised later in SALT II and again in the INF negotiations and is not about to disappear from the agenda of the arms control talks. Besides, it is remarkable that the concerns about conventional asymmetries became acute when the process of nuclear arms limitations started in earnest, and that the green light to the CSCE and the MBFR talks has been given after the signature of the first SALT agreements in Moscow. In his speech at the International Institute for Strategic Studies in October 1977, Chancellor Helmut Schmidt drew attention to the strategic implications of a SALT II agreement which was supposed to neutralize the central systems and underline the regional imbalances at the theatre nuclear (the grey zone) and conventional levels. Finally, the treaty on the elimination of intermediate and short-range land-based missiles (December 1987) gave an impetus to the discussions on the ways and means to stabilize the conventional balance at a lower level of armaments.

Thus, the reductions of nuclear weapons which are viewed in the West as the main instrument of war prevention seem to imply a parallel effort to reduce the incentive to attack with conventional weapons. Conversely, the correction of the conventional asymmetries seem to allow broader limitations and reduction in the nuclear arsenals. This point has been made by the NATO countries at the Reykjavik meeting in June 1987 in connection with the so-called "third zero option". It was stated that a "coherent and comprehensive concept of disarmament and arms control would include - in conjunction with the establishment of a conventional balance and the global elimination of chemical weapons - tangible and verifiable reductions of American and Soviet land-based nuclear missile systems of shorter range, leading to equal ceilings". If the existence of imbalances at the conventional and chemical levels is perceived as an obstacle to further reduction of nuclear systems, the final aim should not be a total denuclearization of the European continent but the reduction to equal levels of the nuclear weapons deemed necessary to keep the balance and to prevent the occurrence of war as long as their elimination cannot be safely envisaged.

The MBFR talks

At the beginning of the MBFR talks the positions of NATO and Warsaw Pact states were in sharp contrast. On the Western side, the emphasis was put on the withdrawal of foreign troops and on asymmetrical reductions of conventional armed forces, the main constraints being put on the Soviet armoured and mechanized units based in the GDR. The reductions of national forces had to occur in a second stage with the aim of reaching equal levels of troops (700 000 men). The nuclear forces were excluded from the negotiation. On the Eastern side, the reductions had to be proportional or equal, as an approximate balance was supposed to exist between the two camps; national and foreign troops were taken into account at every stage of the reductions. Finally, conventional and nuclear components were put together and treated in combination as part of the global equilibrium. In 1975, the United States made a concession to the Soviet Union in accepting to trade 1000 nuclear warheads and 90 delivery vehicles (aircraft and missiles) for a Soviet tank army. This proposal, known in the NATO circles as the "third option", raised controversies among the allies, and the Soviet Union took advantage of the opportunity to insist on the inclusion of nuclear weapons in the Vienna talks. Eventually, the issue was settled in another way: the NATO countries decided to remove 1000 warheads in connection with the modernization of the long-range theatre nuclear forces (double track decision of 12 December 1979), and the Warsaw Pact states announced the withdrawal of two armoured divisions from the GDR. Later, a consensus emerged on the provisional exclusion of nuclear weapons from the MBFR talks, but the question has been raised later in another forum.

The Road to Conventional Stability Talks

When the French President, Valery Giscard d'Estaing, in his speech before the General Assembly of the United Nations, in May 1978, proposed the convening of a conference on disarmament in Europe, it was understood that nuclear weapons would not be the object of negotiations. The argument was that these weapons needed special treatment and that it was wrong to isolate the nuclear forces on the European theatre, since the strategic weapons could also be used in Europe on the same grounds as the tactical weapons. Considering the conventional imbalances, the priority should be given to the reduction of the major ground and air forces with offensive capacities, and the process should start with militarily significant and verifiable confidence-building measures applicable in a geographical zone stretching from the Atlantic to the Urals. This approach raised many objections but finally the 35 signatories of the Final Act of Helsinki accepted the French scheme of a Conference on Confidence and Security-Building Measures and Disarmament in Europe (CDE). The mandate of the CDE was adopted in Madrid (September 1983) and negotiations started in Stockholm in January 1984. In September 1986, a first agreement on confidence and security-building measures (CSBM) was reached and the question of a follow-up conference devoted to disarmament has been discussed in Vienna since the beginning of 1987.

Without going into the intricacies of the present discussions, it seems that the NATO and Warsaw Pact states disagree on the agenda of the conventional stability talks (CST). Whereas the Western countries consider that only conventional forces are eligible for reductions, the Eastern part is inclined to make a case for the inclusion of tactical nuclear weapons. Unless one can conceive a compromise formula to bridge these differences, before the end of the Vienna follow-up conference of the CSCE, it is likely that the issue will be raised in the future and will remain a bone of contention between East and West in the European security debate. Another sensitive point is the existence of dual purpose weapons able to deliver nuclear as well as conventional munitions, and the difficulty to verify the elimination of nuclear warheads assigned to these systems. In this respect, there are different views not only between the leaders of the two alliances but also within the Atlantic community itself. The Germans are concerned about the threat emanating from the battlefield nuclear weapons should deterrence fail, and there is a wide consensus among the political parties, be they conservative or social-democrats, to reduce the risk inherent to the deployment of these kind of weapons in central Europe. The other allies, and especially the nuclear powers, are more concerned about the risks of a denuclearization of Europe, and fear that the inclusion of the dual purpose weapons in the CST would initiate such a process. For many Western experts nuclear weapons are useful not only to compensate for the perceived superiority of the Warsaw Pact forces but also to keep the credibility of the deterrence relationship. This point was underlined in the statement of the North Atlantic Council (2 March 1988) entitled "Conventional arms control: the way ahead". As the countries of the alliance will remain under the threat of Soviet nuclear forces of varying ranges, only the nuclear element can confront a potential aggressor with an unacceptable risk. Therefore, for the foreseeable future deterrence will continue to require an adequate mix of nuclear as well as conventional forces. This stand seems to imply that stability cannot be realized exclusively with conventional means and that conventional parity has to be coupled with the reduction of nuclear weapons to a level sufficient for deterrent purposes. The Warsaw Pact states have another opinion on this subject, and the communiqué issued on 16 July 1988, at the end of the Warsaw meeting of the political committee, reiterates the view that conventional stability talks should be completed by negotiations on the elimination of tactical nuclear forces.

Conclusion

It is obvious that the conventional and nuclear forces are intertwined and that the military balance in Europe is dependent on these two components. Thus, conventional stability is almost inconceivable without taking into account the nuclear capacities, which are considered by most of the political leaders in the West as a major contribution to prevention of war. This does not mean that the existing force levels should be maintained, and that the nuclear weapons deployed in Europe should be excluded from any negotiation as long as the conventional imbalances are not corrected and the chemical disarmament is not achieved. A link has been established between the reduction to equal ceilings of the short-range land-based missiles and conventional disarmament in Europe, but the practical steps to make progress in this direction have not yet been defined. There are also difficulties in finding an adequate way to regulate the dual-purpose weapons, and attempts are made to settle the issue in connection with the elaboration of a mandate for CST. To overcome these obstacles it is imperative for all the interested parties to make up their minds (the Germans prefer to use the more impressive word **Gesamtkonzept**) on the ways and means to reconcile security and arms control, and to promote a peace order which would allow the self assertion of the national entities, ease the contacts between the two parts of Europe, and which would rest on purely defensive military forces.

Nuclear Forces and Their Relation to Conventional Armaments

Mikhail Milstein

The Nature of the Interrelationship

What is the relationship between nuclear forces and conventional armaments? This is one of the most complicated questions of contemporary military policy, military strategy and of military doctrine as a whole. The nature of this interrelationship and its complexity is determined by a number of factors.

First, as a basic premise, it is necessary to state that under present-day conditions a war of any type between the two great powers, or between the two alliances, may result only in a catastrophe, even in mutual annihilations. Therefore, a war of any type between them is inadmissible - neither nuclear, nor conventional war. It is not by accident that this idea is repeatedly emphasized at all USSR-USA summit meetings and in all joint statements. It was also stressed during the 1988 Moscow Summit. The Joint Statement, which was signed by the two sides during that Summit, states that "they have solemnly reaffirmed their conviction that nuclear war cannot be won and it must never be fought, as well as their determination to prevent any war between the Soviet Union and the United States - be it a nuclear or conventional war"[1]. It may be said in this connection that, in looking into the problem of interrelationship between the nuclear forces and conventional armaments, the main objective should be to define ways of making this interrelationship conducive to the prevention of any war whatsoever, rather than to find new war fighting capabilities.

Secondly, in the past and under present-day conditions, nuclear weapons are considered in the West as the main means of deterrence. This is valid both with respect to the United States and to NATO as a whole. As is well known, NATO continues to adhere to the "flexible response" doctrine, which envisages, as the main element of deterrence, the possibility of a first use of nuclear weapons. The statement issued after the meeting of the NATO Council in Brussels, held in March 1988, contains the following: "The interdependence between nuclear and conventional forces is complex. Imbalance in conventional forces, which favours the Warsaw Treaty Organization, is not the only reason making necessary the presence of nuclear forces in Europe. NATO countries are and will be threatened by Soviet nuclear forces of different range. Though the achievement of parity in conventional forces would promote a strengthening of stability, it is only the nuclear element that can deter an aggressor by a threat of inflicting unacceptable damage. Thus, in the foreseeable future, deterrence will continue to make it necessary to preserve appropriate levels of both nuclear and conventional armaments"[2].

Without discussing the merits or demerits of the assertion concerning the imbalance in conventional forces or concerning a possible aggression on the part of WTO countries, we may only note that this statement confirms the continuing adherence of NATO to the time-worn strategic dogma: nuclear weapons are the mainstay of deterrence and they should be kept in store for a long, long time to come. Meanwhile, modernization of tactical nuclear weapons and of their delivery systems is proceeding at a rapid rate. In particular, their range, accuracy and destructive power continually increase. Obsolete delivery systems and nuclear warheads are being replaced by new units, and even though the total number of nuclear weapon delivery vehicles is being reduced to some extent, the effectiveness and destructive power of nuclear warheads are increasing. Neutron weapon systems are now becoming integrated into the equipment of military units; and we are witnessing the advent of ALCMs, SLCMs and "Stealth" technology. All artillery pieces of 155-mm and 203.2-mm calibre are dual-purpose weapons, because they can be used for firing both nuclear and conventional shells, the range being doubled. Other qualitative changes are also underway, actually bringing all of us closer to a new spiral of the arms race, rather than strengthening stability.

Speaking about nuclear weapons in Europe and their relation to conventional armaments, one cannot but emphasize that the exchange of ratification instruments, which took place during the Moscow Summit, has put into force the INF Treaty on the elimination of Soviet and American medium- and shorter-range missiles signed in December 1987. Historically, this is the first international agreement which envisages the elimination of two classes of nuclear armaments of the USSR and the USA. On the one hand, the Treaty marks a practical beginning for constructing a world without nuclear weapons and, on the other hand, it establishes new standards and approaches for arms limitation and reduction. However, a campaign in favour of so-called "compensation" has been gaining momentum in the West, the proponents claiming that it is necessary for the West to replace the forces subject to destruction. It is not clear yet how they propose to actually carry out this "compensation". One thing is quite clear, however: any "compensation" can only result in a new round of the arms race and thus will undermine the significance of the INF Treaty, as well as the possibility of continuing the construction of a world free of nuclear weapons.

Thirdly, discussing the interrelationship between nuclear forces and conventional armaments, one has to keep in mind that nuclear weapons in certain forms have long ago become an integral and intrinsic part of conventional armaments. Atomic field artillery, short-range missiles, nuclear mines, dual-purpose weapons, tactical nuclear-capable aircraft - all these have already become integrated into the equipment of the so-called general-purpose forces. On the other hand, the continuing improvement of conventional armaments brings the destructive properties of conventional weapons progressively closer to those of nuclear weapons. Greater fire-power of conventional weapons is achieved through making qualitatively new weapon systems, as well as through developing highly efficient reconnaissance-attack complexes, capable of delivering powerful strikes far beyond the lines. Troops are supplied with large quantities of multiple-launch rocket systems (MLRS). The salvo of one such 12-tube rocket system may equal in its destructive power that of three 203.3-mm howitzer batallions. Research and development is underway on a cluster warhead for the MLRS with several homing submunitions[3]. On the whole, as already stated, conventional weapon systems are being re-equipped on a very large scale involving still more destructive long-range and high-accuracy weapon systems.

The Statement issued at the conclusion of the meeting of the NATO Council, in March 1988, reaffirms that the "armed forces (i.e. the armed forces of NATO) are intended exclusively for the purpose of preventing war and for self-defense..." And the statement goes on to say that "our ability to prevent any war - nuclear or conventional - from occurring is based on our capacity and ability to deter an aggression in any form"[4]. However, judging by the nature of this document, nuclear weapons continue to be the centrepiece of deterrence. The document is silent or vague on the role of conventional forces and on their relation to nuclear forces.

Fourthly, the connection between nuclear forces and conventional armaments can be seen both in attempts to solve the problem of radical reductions of nuclear weapons and in the search for deep cuts in conventional forces. To a certain extent, these two problems are interconnected, and each can affect a successful outcome of the other. A radical reduction of the strategic offensive nuclear forces would, undoubtedly, affect in a positive way the possibility of achieving an agreement on deep cuts of conventional forces in conditions preserving and strengthening stability and maintaining parity at a lower level of military confrontation. These reductions also would necessarily change the interrelationship of the roles of conventional and nuclear forces as the two key elements for strengthening stability and security and for preventing war; that is their roles in what is meant today by deterrence. In this case, the role of conventional forces will no doubt grow.

The Concept of Deterrence

The interrelationship between nuclear forces and conventional armaments in the context of the problem of security cannot be correctly and thoroughly understood and evaluated without a critical analysis, even a brief one, of the concept of deterrence. In its modern interpretation, the concept of deterrence based on nuclear weapons originated simultaneously with the advent of nuclear weapons and was developed in parallel with the development of these weapons. It envisaged in the past, and it also envisages at present, the possibility of using nuclear weapons, irrespective of how noble the underlying motivations could be. It is based on the premise that nuclear weapons are the principal means of deterring an aggression and, therefore, the principal guarantor of security. And there are absolutely no differences at all about what is actually meant: a threat of an unacceptable retaliatory nuclear strike or a first use of nuclear weapons in Europe as the last option for repelling an aggression of conventional forces from the East. What is implied is real use or a real threat of use of nuclear weapons, without which deterrence becomes an empty word, a mere fiction. One need not possess a lot of strategic perspicacity to see that in both cases the idea is to deliver nuclear strikes at targets located at the territory of the other side. For it can hardly be assumed that somebody would plan inflicting nuclear strikes on his own territory even if it were invaded by the enemy.

Many years have gone by since this concept was adopted, and many things have changed. The size of nuclear weapons stockpiles has increased tremendously and, as is said today, a suicide potential has been accumulated. Views and ideas as to the probable consequences of military use of nuclear weapons have also changed, as has the perception of what would happen to the world if a massive nuclear exchange ever took place. Most people believe that such an exchange would result in a catastrophe and may possibly lead to the complete destruction of our civilization. Views on the feasibility of a limited nuclear war have also undergone noticeable changes, with the result that few people admit nowadays that such a war is possible at all, with more people believing that inevitably a limited nuclear war would become an all-out war, with all the cataclysmic consequences. Opinions as to possible uses of nuclear weapons as a means of implementing the "policy of strength" have also changed. In a word, nuclear weapons are gradually becoming (or, maybe, have already become) dead weapons that cannot be used at all - neither militarily, nor politically.

The time-worn dogmas and the blind beliefs in the divine strength of nuclear weapons are too strong, however, to make some people admit this truth simply and unequivocally and to start looking, on a joint basis, for practical and safer alternative ways of strengthening stability, preventing establishing a system of lasting security both for one's own country and for the world as a whole. But let us go back to the concept of deterrence, because so far nobody has abandoned it. It presupposes the availability of such quantities of nuclear forces, with their tactical and combat specifications, plans for their military use, and with their maintenance in a state of combat-readiness that the other side will not harbour any doubts as to the consequences of a retaliatory strike which would not be less destructive than the first strike. It is these doubts that, conceivably, must make the other side refrain from inflicting a first strike. In so far as the first use of nuclear weapons is concerned, it is implied that the first strike - on targets located on the territory of the Eastern bloc countries - should be so devastating that it would make the enemy (i.e. the "aggressor") discontinue all offensive operations and immediately start asking for truce on conditions which would be to the advantage of the Western bloc countries, that is to the side which would be the first to use nuclear weapons. The role of conventional armed forces, as well as their relation to nuclear forces in the course of a practical implementation of the concept of deterrence, remains vague and is never explicitly formulated.

What conclusions can be drawn from the above? Successful implementation of the concept of deterrence is intrinsically connected with intimidation and taking advantage of a fear of nuclear weapons. The concept of deterrence cannot be effectively implemented without a permanent build-up and qualitative improvement of nuclear weapons. At the same time, continued accumulation of nuclear arsenals, their incessant qualitative improvement, and the drastic reduction in the flight-time to their targets of nuclear delivery-vehicles, and hence in the time for evaluation of the military situation and for rational decision-making, increase the danger of an unauthorized or accidental outbreak of war. Essentially, "nuclear deterrence" cannot exist apart from ever-growing arsenals of conventional weapons and their constant qualitative improvement. The most fundamental premise of nuclear deterrence is perpetuation of nuclear weapons, which creates insurmountable difficulties in the way of radical reductions of both conventional and nuclear weapons. A dangerous link is created between nuclear forces and conventional armaments, whereby even a minor conflict may become a fuse triggering an all-out nuclear war. Moreover, it should not be forgotten that nuclear deterrence by implication exerts a negative influence on the taking of effective steps in the field of nuclear non-proliferation. This in turn creates a permament threat of the enlargement of the "nuclear club", particularly in the world's hotbeds of tension.

Thus we see that under conditions of a permanent threat of mutual suicide posed by nuclear weapons to the further development of the world's political climate in a positive direction and to the beginning of a process of eliminating nuclear weapons, the concept of nuclear deterrence, which is based on a threat of the use of nuclear weapons, is not only an extremely dangerous concept but also a hopelessly obsolete one. This is why this concept cannot serve as a guarantee of security in the future.

The Elimination of Nuclear Weapons

To be sure, it is not a simple matter to suddenly abandon the concept of deterrence, which has been in existence for over 40 years. In a way, it has won a mandate or permanent existence; besides, it has been providing a suitable and outwardly convenient key of solving the problem of security.

What the world needs now is to give renewed attention to the possibility of solving the problem of security and, more importantly, to overcome not only conceptual but also psychological barriers, that is those which are the most difficult ones to overcome.

Any ideas about totally eliminating nuclear weapons brings about an allergic reaction in the West, thus enhancing fears and concerns with respect to stability and security, including problems of the interrelationship between nuclear forces and conventional armaments. In 1986 the Soviet Union, as is well-known, put forward a concrete programme for the complete elimination of nuclear weapons during a 15-year period, to run in parallel with the liquidation of other types of weapons of mass destruction and a radical reduction in conventional weapons. Incidentally, that Soviet programme mapped out prospects for conceptual changes from nuclear deterrence to other, more effective ways of strengthening security.

The entire world, it would seem, should have welcomed these proposals, because their implementation would be in the interests of everybody, without giving any advantages to anyone but bringing freedom from the Damocles sword of the nuclear threat which has for too long been hanging over the world. Few in numbers, however, were the voices of influential public personalities, scientists and governments of Western countries who supported these Soviet proposals. The arguments put up to cast aspersions on the proposals are legion. It was said, for example, that it would be impossible to eliminate nuclear weapons since it is impossible to eliminate the scientific knowledge and know-how involved in their manufacture. The proposals were branded as utopian, and it was stated that complete elimination of nuclear weapons would be to the advantage of the Soviet Union because of its alleged superiority in conventional forces. In all this it was never forgotten, and actually it was even stressed, that nuclear weapons were the guarantor of security and that they had heretofore deterred and would continue to deter the world from a nuclear war. In the heat of the argument, whether by omission or by commission, it was never mentioned that negotiations on the prohibition and destruction of chemical weapons had been underway; that biological weapons had already been banned (scientific knowledge and know-how being no obstacle to it); that the Soviet Union was not proposing an immediate elimination of nuclear weapons but rather suggesting a span of 15 years, which is not a brief period of time; and finally, that the Soviet Union had simultaneously put forward a programme for the reduction of conventional armaments and for the establishment of parity at much lower levels. These Western "arguments", on the whole, mirror very accurately the fact that the West was both psychologically and strategically unprepared to abandon nuclear weapons and to commence their complete elimination.

Security by Non-Nuclear Means

What is it which can replace the concept of deterrence, and what must be the relationship between nuclear weapons and conventional armaments in the context of solving the problem of security in the future? And, in general, can security be reliably assured by non-nuclear means so that one's own and common security would not be jeopardized? The Soviet Union is convinced that it is possible to ensure security by non-nuclear means on the basis of the principles of sufficiency. The process is a complicated one, and not only that. For to a certain extent it is a long process. As stated earlier, the period covered by the Soviet programme for the elimination of nuclear weapons is 15 years. If the disarmament process continues to be extended and without losing any momentum, then this period of 15 years will be characterized, **inter alia**, by an intensive search for a lasting and reliable structure of security using different approaches and a different qualitative level in comparison with what we know today, because one cannot proceed on the way toward complete elimination of nuclear weapons without taking into consideration other important elements of security and, above all, the role of political factors and of the evolution of the armed forces and conventional armaments.

It seems that the role of political means in ensuring security will be enhanced. It also seems that much effort will be devoted to the extension and practical implementation of a wide spectrum of confidence-building measures between states. It is only enhanced confidence which will provide a key to the door for implementing new approaches and new solutions and which would relate to present-day requirements. We can expect, before the time comes when once and for all there will be no more nuclear weapons in the arsenals of the nuclear powers, that the role of nuclear forces in ensuring security will gradually diminish, while that of conventional forces and armaments will, on the contrary, augment.

The way towards non-nuclear assurance of security passes through a replacement of nuclear strategy by a non-nuclear one, and not only by a non-nuclear strategy, but by a strictly defensive non-nuclear strategy, by one which is based on the principle of sufficiency. The gist of the matter is mutual recognition and mutual confidence in that both sides have **enough** strength only to provide for their own security and to repel aggression by self-defence, but **not enough** by far to conduct offensive operations. In this case, by mutual agreement, the armed forces of both sides would have such a structure of military equipment and armaments, such deployments and strategic orientations, as to absolutely rule out, from the viewpoint of military technology, the possibility of a sudden attack or offensive operations. A mutually agreed reduction of military potential will also be required, the level of such reduction being such, that neither side, while being able to ensure its own security, would have sufficient strength or material means to carry out offensive operations.

The Warsaw Treaty Organization member-states have declared their adoption of a military doctrine of a strictly defensive character. They have addressed an appeal to NATO member-states to carry out a review of their military doctrines with a view to bringing military theory and practice into agreement with the requirements of defensive strategy and the principles of sufficiency for defence.

It is difficult to overestimate in this connection the importance of negotiations on conventional forces and armaments, particularly in Europe, the objective of which would be to solve the problem of security and stability at a lower level of military confrontation, and to eliminate, on the basis of drastic cuts, the imbalances and asymmetries.

As for other measures for easing the military confrontation in Europe, with its highest concentration of troops and armaments, great importance should be attributed to the disengagement of confronting troops, and the establishment of nuclear- and chemical weapon-free corridors and zones.

These are some ideas concerning the interrelationship between nuclear forces and conventional armaments. The nuclear component of strategy is the most agressive catalyst of the arms race, and this is exactly why its role must gradually decrease, until finally it can be done away with, once and for all. Conventional forces must be reduced to the threshold of sufficiency. A new situation requires new approaches, which have to be based on the most fundamental premise - that any arguments must take place under the conditions of peaceful coexistence and must be solved by political means.

This does not at all mean that divergences between the USA and the USSR, or between NATO and WTO, could as a result be ruled out, particularly so far as the plans and intentions of the two sides are concerned. But it is important, however, that these divergences should not lead to a military confrontation or to increased tension. It is important to recognize firmly that no contradictions exist which would fatally doom the USSR and the USA to a confrontation, to say nothing of a war. During the nuclear era and the space age the world has become much too fragile for war and a policy of strength.

References

1. Pravda, 2 June 1988.

2. "Non-Offensive Defense", No.9, May 1988, p.7.

3. For details cf. "Whence the threat to peace", Moscow: Military Publishing House, 1988, p. 48.

4. "Non-Offensive Defense", No.9, May 1988, p.7.

Restructuring Conventional Forces in Defensive Modes

Peter Deak

The approach to this theme is extraordinarily difficult. The main reason is that the completely theoretical approach does not give anything, and the concrete approach requires, even by a military expert, raising elements that intrude into details of the armed forces of certain countries. The intention of this paper is to remain within the area between abstractness and concreteness, and not to touch by assumptions the concrete elements of the armed forces of either side.

To begin with, it is very important to understand precisely the phrase "conventional forces". Without a qualifying definition the restructuring may allow a numerical decrease of offensive elements, within a certain weapon's range, but at the same time increasing their capacity; the modification into a seemingly defensive feature would thus essentially result in an offensive feature.

According to military science, conventional forces are non-mass-destruction, but suitable to launch. In this sense, the first step of restructuring conventional forces into a defensive direction is back-structuring, that is re-establishing the conservative character of forces, thus depriving them of all the means capable to operate nuclear or chemical weapons, target acquisition, controlling and transportation systems. Thus, the first step would be to re-establish the conservative character of the land, air and maritime forces, so that the double purpose, or multipurpose means would be part of those to be pulled out.

What do I mean by conventional forces? In one sense I mean the structures before the proliferation of nuclear weapons. In another sense, those weapons introduced because of constraints upon nuclear armaments. The latter have reality, since the military-technical development of the last 40 years cannot be erased; the military technology of the developed non-nuclear armies of the neutral countries and in the Third World must be accepted.

By no means can we consider, however, as parts of conventional forces the weapons of an offensive character to be introduced or developed as a result of the limitations of nuclear weapons (basically the SALT, ABM, and INF) agreements whose effectiveness and importance is not yet foreseeable. Ignoring this would result in a tendency towards subsequent re-armament. It results not in "re-" but "OVER-STRUCTURING".

The attribute "conventional" supposes in the military sense that, in general, the weapon cannot be used without the deployment of forces and supply services. The "conventional" weapon has no independent strategic strike capability. It also means that its range, speed and precision is so limited that it does not restrict the possibility of counterattack significantly, and it does not possess disarming capability. The next criterion is that it includes only the classic offensive weapons, such as bombers, tanks and other weapons. It is also important to note that it needs the presence of humans. The basic purpose of restructuring in a defensive direction is twofold. First, the restructuring must be in a certain direction, e.g., organization into a "low" level. Second, the reduction - and not replacement - of certain components, to ensure a higher proportion of defensive elements, of defensive, reconnaissance, preventive capabilities.

When speaking about the structure of armed forces it should be noted that the man-machine relationship is hidden in the modern armed organization, which includes input, output, subsystems, system elements, and the definite proportions of the individual elements. Military structures work by specific mechanism. They are human systems in which the alteration of one of the elements of the armed services results in a qualitative change in the whole structure.

Parts of the military structures are their special dislocation. If we examine them from offensive and defensive aspects, they involve the vicinity of the whole unit to the line of attack, to the marching order, the relationship to the operational directions suitable to attack or defence, and to deployment for battle. The inner grouping of the given unit is, however, also important, namely, whether its dislocation meets or does not meet the shaping-up of the war offensive or defensive activity; its subunits, how much are they near to each other, what the concentration of the organization is, because a high degree of this implies a capability to attack, while its dispersion means the ability to protect. In a general sense forward deployment expresses an ability to attack, whilst deployment on the boundary or adjacent to it infers a defensive intention.

The offensive or defensive character of the structure is mainly expressed by the proportions of the pure offensive and defensive elements. Definition of "pure offensive" or "defensive" means is rather difficult. Heavy tanks in mass are of offensive capability and character. However, in small units, or as single ones, in several historic cases they constituted the roots of defence. Neutral countries with a defence structure built on defensive doctrine have fortified territorial defence with obsolete tanks or turrets. An opposite example is the mine. Its conventional version is to block the way of the attacker, it is a passive element of fortification. Nevertheless, projecting them into enemy territory by the use of cannons, missiles, or airborne means, they can be used as offensive weapons.

Let us try to clarify or categorize weapons from the point of view of offensive or defensive capability. Weapons of typically offensive character are those which - due to their individual capabilities, complexity, integrated capability - have enhanced range, precision and destructive power. Such is, as an example, the self-guided missile with a range of 100 - 200 km, equipped with conventional warhead. It is, in itself, if used against attacking armoured or air targets, a defensive, protective weapon. It can accomplish defensive missions aboard fighter aircraft or helicopters. If this missile with the range of 100 km is mounted on a bomber, then the bomber's range must be added to the missile's one, so the total effective distance is accordingly increased. Its precision is enhanced by the controlling and correction devices aboard the aircraft; its destruction capability is increased by the fact that the air-launched missiles are applied not one-by-one but in great numbers to cover the target area. If this aircraft is based on a carrier, manoeuvreing in an ocean area, the original range is again multiplied due to the forward position of the ship. The massive application is also increased, because complete air units are based aboard, and also the preciseness, because a fully developed land-airsea C^3-I system guides the weapons. All these are further increased (mainly the delivery range) if the aircraft-carrier is located at a distance from the main base.

What are the points of view by which it can be stated whether a weapon or device is of an offensive, defensive, mixed character, or is insignificant equipment? In my judgement weapons of offensive character are those which:

a) exceed the force necessary to detect and destroy them;

b) are deployed on hostile territory;

c) are designed to destroy forces, defensive techniques and equipment on hostile territory

d) are capable of occupying and holding territories, bridge-heads, etc. needed to undertake further offensive activities, such as:

  - missile complexes of long range, high velocity, high precision navigation;

  - tanks equipped with offensive weaponry, high travelling speed, with a 150 - 200 km action radius;

  - sea and air lift systems that are capable of transporting manpower equipped with much combat weaponry at a 200 - 200 km distance;

  - pontoons capable of rapidly bridging wide waters, for transportation of heavy weaponry;

  - long-range artillery, beyond the range of 40 km;

  - powerful radio jamming equipment that can block energy plants, water plants, civil defence and catastrophe-preventing radio networks stationed beyond the combat zone;

  - short-range weapon systems, robots (drones, robot-like land vehicles, devices flying at low level) that are intended to break through the homeland air defence and stable land civil protection systems;

  - all transportation devices carrying, transporting and delivering the aforementioned weapons, surface ships, submarines, aircraft-carriers, and flying devices;

  - diversion forces.

Of course, the list is not complete. A precise definition requires an expert group composed of men highly trained in military and technical matters. A prerequisite of all this is an adequate exchange of information and agreement on its authenticity.

Typically defensive means are, in my view, the following:

a) they are stable, not removable, standard systems, the time required for their eventual prepositioning exceeding that of prepositioning the means needed to counteract them;

b) they carry out their activities on their own territory, and their control system cannot be redeployed;

c) their range does not reach the limits and depth of the so-called defensive zones;

d) they form obstacles to attack;

e) they can jam obvious airborne attacks, other possibilities excluded;

The following weapons can be regarded as defensive:

- stable systems ("built-in elements") of fortification, forward deployed automatically or man-controlled mine fields, trap fields, artificial water obstacles and devices permitting destruction, respectively;

- river mines anchored in the water;

- detection devices placed in the region of seashore, and fixed mines controlled by them;

- coast defence fortifications including fixed medium range artillery pieces;

- coast defence naval ships whose action radius does not exceed the range of the shipboard artillery and missile systems;

- combat-ready, coast-based helicopter-carrier groups, including submarines whose range does not exceed 360 km, together with the shipborne antisubmarine helicopters;

- short range interceptor and reconnaissance aircraft of the marine air force;

- all the elements of the homeland air defence, capable of operating only on home territory (missiles, short range typical fighter aircraft and their radar systems);

- antitank forces and equipment of land troops that carry out their task basically on conventional built-in bases or firing positions (i.e. carrier platforms);

- radio jamming equipment that is intended to prevent guided missiles from reaching their targets.

Mixed purpose conventional forces are as follows:

- the intelligence forces and equipment (since they are capable both of determining the danger of a hostile attack and the possibilities for one's own attack, respectively);

- medium and short range artillery;

- mobile antitank weapons, light combat vehicles;

- helicopter gunships capable either of antitank protection or short distance strike;

- tactical (i.e. short range) landing forces and unarmed aircraft transporting them;

- (disputable) close air support aircraft and, finally,

- prepared bridge elements allowing traffic in several directions, on one's own territory.

The most characteristic and global factor of restructuring is the alteration of proportions of armed services. This question falls in the strategic category and is not limited to conventional forces. Let me detail some points.

1. There are the given homeland air defence systems. The prohibition of their maintenance or development, or their reduction are questions beyond any disarmament principles. There is no need to introduce alteration or restructuring in the sphere of homeland air defence. In general, the homeland defence systems must be considered a defensive armed service, whose basic intention is to protect the territory, industrial, economic regions, infrastructure, population and other values. It must be taken into consideration that - due to their dispersion - they are at the same time able to defend the military command posts and unit groups ("goose eggs"). Concentration of these has to be eliminated by confidence-building measures, not by restructuring. As a restraint in the case of air defence missiles, none must be developed that exceed long range artillery; in the case of fighter aircraft none must be developed that can reach the range of bombers, nor must their missiles reach the range of the long range artillery.

2. Strategic missiles have nuclear warheads, and are beyond the scope of this paper.

3. As far as the navy is concerned, aircraft carriers must be withdrawn from the weapon systems, and their modernization stopped. Some part of this problem area also falls into the nuclear circle. In an adjacent sphere: large surface ships are in any case very vulnerable, so withdrawal of aircraft carriers would result in the total reorganization of the navy. Among the other surface naval groups those protecting the long sea communication lines are not offensive in character, because their mission is defensive: to prevent and disturb long distance prepositioned attack. Naval groups of offensive character are those intended to attack, destroy and/or to land on shores, ports, and energy plants of continents. Floating docks are also offensive. Among the ship types that must be limited or pulled out of service are landing craft that are capable of disembarking great manpower and heavy weaponry and floating, deployable ports and docks. Also, dual purpose weapons must be removed from ships.

4. An understanding of restructuring requires establishing a basis for comparison of various combat equipment. Defining the proportion between defensive and the offensive weaponry involves basically numeric values and numbers of personnel. Corrective factors such as mobility, manoeuvreing indices, speed, precision, protection, river crossing capability and fire power must also be examined.

The proportion between combat and support forces is important as well as between the quantity of operative combat material and the quantity of munition and fuel. The larger the storage, the greater is the offensive capability, especially offensive elements that are not stored in the hinterland, but are mobile. As an initial step, for confidence-building, one should consider reducing the storage needed to operate conventional offensive weaponry or its redeployment at long distances, including verification.

5. There are some detailed questions on the restructuring of land troops. The most typical are the infantry, rifle, armoured, Panzergrenadier, tank regiments, brigades, divisions and their equivalents. In general, in a modern organization, the proportion of tanks to other ligher armoured vehicles is between 1 to 4, and 1 to 5; the proportion of the total amount of armoured equipment capable of attack to other combat, transport etc. equipment is 1 to 4, and 1 to 5, depending whether it is armoured, mechanized infantry, or Panzergrenadier units.

An obvious solution would be to disband armoured organizations above the battalion level; removing some part of heavy weaponry -, if there is no mutual troop reduction - or degrading them into infantry or defensive support organizations. It would be an ideal limit, at the same time realistic, if for subordinated units, where offensive weapons remain, their proportion of manpower and weaponry could be limited to no more than 1 to 10.

Experience suggests that each antitank weapon is - in general - capable of destroying mobile hostile armored targets, thus their proportion generally is 1 to 3 in comparison to the enemy. The superiority of defence would be expressed in an appropriate way if the proportion of antitank to attacking armoured weapons were 1 to 1.

As an initial step of restructuring, the artillery organized on the basis of its calibre could be transfered to the next higher level unit. Thus, the present battalion artillery would go to the regiment/brigade, the brigade artillery and missiles to the division, and from this to the corps and then to the armies. Long range artillery and missiles above army level should be withdrawn. Dual purpose weapons, mainly short range missiles and large calibre nuclear capable artillery, must be disbanded and pulled out of the system, and the missiles destroyed in a verifiable way. The artillery pieces must be dismantled and not be stored as organic weapon systems. This can be verified at the bases.

The bomber and fighter bomber aircraft must be removed from the organization of the corps and army, although a certain part of them may be kept within the army group and higher organizations. Their air bases must be at a distance three or four times further than their range, and their fuel must not be allowed at intermediate support air bases, nor may they be stored at such bases.

The ability to maintain and repair heavy land weaponry must be degraded in a significant way.

6. Taking the long process all in all, a time table of restructuring may consist of several stages. One possible sequence is as follows:

a) Decreasing, pulling out, displacement of necessary operating equipment.

b) From among the weapons characterized as offensive, pulling out of the system those qualified by the given country as obsolete, or by mutual agreement on numeric and comparative equivalents, by way of mechanical destruction, without replacement. This process results in a linear decrease of offensive elements.

c) The military verified dismantling of obsolete weapons among those qualified as offensive; accounting, storing their equipment, devices applicable for defensive systems, or connecting them with defensive zones, their application as protective weapons. Mechanical destruction of the carrying platforms, or their guaranteed civil economic usage. This process is not a linear decrease of offensive elements, because the complementary devices are enhancing factors.

d) Withdrawal and prohibition of basically offensive units (e.g. armoured major units, assault regiments, aircraft-carrier groups) from a given region, with assignment of entry points/lines. In a certain sense this is an interim solution, associated with the putting of these units into strategic reserve, with a guarantee that replacement of the out-of-date items must not take place.

Finally, I should like to make four important points.

Firstly, in no circumstances must restructuring lead to significant expense, because this may result in the danger of a new defence arms race. One must think primarily of the existing arms categories.

Secondly, the basic condition of restructuring is withdrawal. Pulling out of foreign troops mutually, behind one's own boundaries (included in several proposals) results from the outset in structural reorganization, since the great powers possess the most offensive units and weapon systems.

Thirdly, parallel with the withdrawal of heavy weaponry related to limited manpower cuts there is a possibility - mainly in the boundary regions - of establishing reliable antitank zones, including light infantry combined with militia and reinforced border guards.

Fourthly, as a compromise, either on a unilateral or a bilateral basis, putting and maintaining certain offensive weapons technically in a temporary disabled state to attack. This may result in structural reorganization within the active zones, because this state can continuously be verified.

Interaction Between Nuclear and Conventional Arms-Control Measures (Report on a Pugwash Workshop, June 1988)

John Holdren

It seemed appropriate in the aftermath of the achievement of the INF Agreement to examine, in a Pugwash expert workshop, the prospects for further arms-control agreements in both the nuclear and the conventional-forces spheres, with special emphasis on the interactions between the two. Combining the topics in this way now seems imperative for at least three reasons. First, the political momentum, public expectations, and negotiating precedents established by the INF Agreement enhance the possibilities for conventional forces arms control as well as for further nuclear reductions. Second, some of the arms-control problems that remain to be solved are similar enough in the nuclear and conventional spheres that it can be advantageous for specialists in the two spheres to compare notes. Finally, and perhaps most importantly, nuclear forces and conventional forces are directly linked by aspects of doctrine, strategy, force structure, and arms-race dynamics in ways that make it militarily and politically impractical to achieve really major reductions or restructuring in either category without reference to what is happening in the other.

Some of the most important nuclear-conventional linkages arise from the perception of many people that the conventional-forces confrontation in Europe would be intolerably dangerous except for the extra caution induced in military and political leaders by the presence of nuclear weapons. This view finds concrete embodiment in NATO's doctrine of extended deterrence - the explicit threat to be the first to use nuclear weapons if that seems necessary to stop a conventional attack. The view that nuclear weapons are needed to stabilize the conventional-forces confrontation has arguably been a major source of pressure for "modernization" of nuclear armaments over the years, with NATO seeking ways to make credible that it could and would use nuclear weapons if necessary, and the Warsaw Treaty Organization seeking ways to assure that escalation to nuclear conflict would bring no advantage to NATO.

The actual result of four decades of this competition in nuclear weaponry is that any use at all of nuclear weapons in conflict is likely to lead to total disaster for everyone. It is hard to deny that this circumstance does produce some additional incentive for leaders on both sides to avoid conflict, beyond that provided by the expectation that even a strictly conventional war in Europe would be unimaginably destructive. The more controversial question is whether the extra caution induced by nuclear weapons is worth the extra costs and risks these weapons bring, including the danger of accidental or inadvertent nuclear war, and the contribution of provocative nuclear-weapons postures to East-West tensions, and the diversion of attention and resources from other military and non-military problems.

Although the participants in the Workshop held differing views on whether a nuclear contribution to deterrence of conventional conflict in Europe is really needed at all, or is worth its costs and risks, there was virtually unanimous agreement that the nuclear-deterrent effect that some think desirable as extra insurance against a conventional conflict in Europe **could be amply provided by nuclear forces far smaller and less threatening than those that exist today.** This conclusion holds not only for nuclear weapons deployed in European countries - weapons that some think could and should be eliminated altogether - but also for those deployed in the United States and the Soviet Union; and it holds for the most basic nuclear-weapons function of deterring **nuclear** attack as well as for the "extended deterrent" role.

It was also emphasized and generally agreed that nuclear arms control should focus not so much on reducing the aggregate numbers of these weapons as on eliminating or strictly limiting those specific types of weapons that most contribute to crisis instability and arms-race instability. Crisis instability is associated with incentives for and fear of pre-emptive strikes, and the nuclear-weapons characteristics that aggravate this problem include MIRVing, high accuracy, and vulnerability to pre-emptive destruction by the other side. Arms-race instability comes from mutually reinforcing incentives to expand the types and numbers of nuclear weapons, and is associated with nuclear weapons whose numbers are difficult to verify, as well as with trying to give nuclear weapons functions beyond deterrence of nuclear attack. **The kinds of nuclear weapons that should have highest priority for deep reductions or elimination according to these criteria include MIRVed ICBMs in fixed silos, nuclear artillery shells, and sea-launched Cruise missiles.**

Workshop participants were in general agreement that, from the military and technical standpoints, and considering the precedents established by the INF Agreement, **there is now great potential for further arms-control agreements that would greatly reduce some of the most dangerous classes of both longer-range and shorter-range nuclear weapons, that would be adequately verifiable, and that would enhance security on all sides.** There is even considerable room for unilateral restraint and reductions - especially with respect to weapons, such as nuclear artillery shells, that clearly threaten their possessors even more than they threaten the adversary. At the same time, it was recognized that from the **political** standpoint there will be considerable reluctance in NATO to agree to any further nuclear-arms reductions that might somehow be thought of as weakening extended deterrence, unless and until the perceived danger from the conventional confrontation is correspondingly reduced.

This nuclear-conventional connection greatly adds to the incentives for conventional arms control that should already be apparent from looking at the conventional confrontation itself. The NATO-WTO conventional confrontation is almost incomprehensibly large (for example, some 70 000 main battle tanks and about 14 million active and reserve ground troops from the Atlantic to the Urals) and almost incomprehensibly expensive (at the equivalent of 600 billion dollars per year). It is arguably no longer just a symptom of underlying political and ideological disagreements, which after all have considerably abated since the confrontation began, but rather is now a primary **contributor** to East-West tensions as well as a primary continuing stimulus to the nuclear-weapons competition.

Much discussion at the Workshop was therefore devoted to the possibilities for building down this increasingly dysfunctional, dangerous, and expensive conventional confrontation. The picture that emerges from this discussion is that **there are tremendous new opportunities for agreement on shrinking and restructuring the conventional confrontation to reduce both its dangers and its costs, and to ameliorate the political obstacles to further nuclear arms reductions.** But there are also major complexities and

difficulties that will require extensive analysis and energetic negotiation if these opportunities are not to be missed.

The opportunities for major progress in conventional arms control arise from a convergence of several important factors: increasing recognition that the conventional confrontation is far out of proportion to any political rationale for it; increasing recognition of its connections both with the causes of the nuclear arms race and with the chances for further nuclear arms control; economic difficulties in many of the countries involved, making it plain that the costs of the conventional confrontation at its current level cannot be sustained; the emergence of increasingly detailed and persuasive analyses of **how** conventional forces could be restructured to provide much greater stability at lower levels; and the important precedents established by the INF Agreement in respect to asymmetric reductions and cooperative verification procedures.

The key idea about the form of the needed restructuring of the conventional forces is that its aim must be to reduce offensive capabilities, including especially the potential for surprise attack, while retaining strong defensive capabilities. The desired "mutual defence dominance" can be accomplished mainly by focusing the reductions on the most offence-capable elements of modern conventional forces (for example, tanks, self-propelled artillery, attack helicopters, and ground-attack aircraft) and by removing the residual stocks of these weapons from forward positions, along with their support infrastructures. The goal of greatly reduced potential for (and fears of) offensive action **cannot** be accomplished merely by pursuing numerical parity in conventional forces.

There is considerable reason for optimism in the steadily increasing credibility of the "defence dominance" idea over the past few years among political and military leaders (brought about in part by the series of Pugwash Workshops on the topic in which analysts from the academic, military, and political communities from East and West have jointly worked out some of the important details). It is now particularly encouraging that these concepts are regularly cited in the speeches of General Secretary Gorbachev and other Warsaw Treaty Organization Heads of State, and that fairly detailed schemes for implementing such ideas are under serious review in a number of defence ministries and general staffs in the West.

Substantial favourable discussion was devoted at our Workshop, in this connection, to the approach to conventional arms control put forward by Mr Gorbachev at the Moscow summit and subsequently described publicly in speeches by Mr Sheveranadze and General Chervov. This approach consists of an initial complete exchange of official data on conventional forces, with resolution of disagreements by on-site inspections, preparatory to negotiating the elimination of the "existing imbalance and discrepancies" in tanks, artillery, helicopters, ground-attack aircraft, and other conventional weapons. A second stage of negotiations under this approach would seek cutbacks of about 500 000 troops, and a third stage would combine further reductions with restructuring in which "the armed forces on each side would be given a defensive character and their offensive nucleus would be dismantled".

This Gorbachev approach is still far from a detailed and formal arms-control proposal, and it would have to be greatly elaborated before one could begin to determine what parts of it NATO would accept; but it already represents an important step forward from the far more limited conventional arms-control approaches that were able to gain endorsement by top leaders previously. The obstacles to reaching a comprehensive and effective agreement on conventional forces are formidable - as 15 years of largely unsuccessful MBFR negotiations in Vienna have demonstrated - but the potential benefits of pushing on to success are huge, and the promise of the new thinking now being focused on the problem is great.

As an illustration of the possibilities that become conceivable with equally bold thinking on the nuclear side of the problem - possibilities that would be made far easier to reach politically if conventional-forces transformations of the sorts described here were actually achieved - the Workshop gave some discussion to a particular scenario for extremely deep reductions by the year 2000 or so. On the NATO side there would be a total of 500 invulnerably based nuclear weapons (200 strategic weapons in the United States, 100 each in France and the United Kingdom, and 100 non-strategic nuclear weapons in Europe altogether), compared to the 25 000 nuclear weapons on the NATO side today; on the Warsaw Treaty Organization side there would also be a total of 500 nuclear weapons, again invulnerably based to assure stability, compared to about 25 000 today.

While the group did not reach agreement on whether something like this 50-fold reduction scenario could really be achieved by the year 2000, or on whether and how such progress could be continued all the way to complete elimination of nuclear weapons everywhere, the fact that such a proposal could now receive serious attention in an East-West working group such as ours is already a most encouraging sign. Such positive vision should serve as an incentive to Pugwash, political leaders, and publics to mount the tremendous further effort that will be required to bring the vision to life.

Part IV

Prevention of Chemical Warfare

International Machinery for Monitoring a Chemical Weapons Convention

Ronald G. Sutherland

Introduction

The problems of chemical weapons have been with us for some 70 years now and the Conference on Disarmament in its many guises has been attempting to deal with the modern threats for almost 20 of them. There are signs of optimism, in the ranks of the negotiators; the end is not here as far as the process is concerned since there is much that is difficult still to do but it is in sight. I would like to stress that time is of the essence and that these problems must be solved soon before the notion of use of chemical weapons becomes commonplace. A weapon system in place will be extremely difficult to remove. The monitoring problems are complex and exacerbated by the fact that we will have to "get them right the first time", as well as have the machinery in place and working by the time the Convention comes into force.

The current draft Convention, CD/782 is 80 pages long and the document contains an additional 32 pages of text which reflect the intercessional aims of the **ad hoc** Committee on Chemical Weapons; we can compare this with an early proposed draft Convention of 1972, CCD/361 which only required five pages of text to outline fourteen articles. The changes we see over this period are evolutionary; as debate on the measures required to control chemical weapons and their production continued, it became increasingly obvious that a more formal approach to international involvement was required and the phrase "international procedures within the framework of the United Nations" was no longer adequate to cover the complexities of the modern chemical industries.

The Consultative Committee and its Subsidiary Organs

In CCD/360, the United States outlined a number of verification considerations which would require national and international involvement and discussed some of the operational considerations together with the possible need for a "consultative body" to administer the international component. This paper was quickly followed in 1973 by two important papers; the Netherlands' paper, CCD/410 which suggested the need for "a plenary Conference, a Board and a Secretariat", and a Japanese contribution, CCD/413 which first specified "an international verification organization". Hence the present day Consultative Committee, Executive Council and Technical Secretariat have been in place in an embryonic form for 14 years. Japan further developed their notion of an "International Verification Agency" in a draft convention, CCD/420 in 1974; its lack of impact implied that the ground was not yet ready for the ideas implicit in a specific international authority to implement a ban on the development, production and stockpiling of chemical weapons. In the UK draft Convention described in CCD/512, the concept of a Consultative Committee was developed more fully and showed the growing consensus for the involvement of all States Parties in the supervision of the proposed Convention. The idea of a Consultative Committee was fully established by 1976 but the actual extent of its powers remains the focus of debate till the present time. Until 1980, there was no clear differentiation between the Consultative Committee and an International Verification Agency. It fell to the Canadian delegation to point out in CD/113 the increasing complexity of chemical weapons control and note that a Consultative Committee would require an International Verification Agency with its own secretariat to carry out inspections and deal with the various kinds of technical, economic and scientific information which would inevitably flow once a Chemical Weapons Convention entered into force. This takes us to the beginning of the contemporary stage of the negotiations where two fundamental ideas converge; the notion that all States Parties must be responsible for the resolution of problems associated with a Chemical Weapons Convention, and that the ramifications of this Convention would require the permanent presence of technical and managerial expertise. The bilateral negotiations of the USA and USSR in 1979 were reported to the CD in CD/48 and included references to a consultative committee and a secretariat which would carry out activities of the former between meetings; a reference which is absent in their 1980 report, CD/112. The reports of intervening negotiations between 1980 and 1983 are very well summarized in CD/416, the report of the last **ad hoc** Working Group. The Consultative Committee and the Technical Secretariat are now well established and the Dutch notion of a "Board" reappears in a new representative group called the Executive Council. Many of the functions of the Consultative Committee are delegated to this body which should be available for relatively quick action both between sessions and also while the Consultative Committee is in session. A hierarchical structure is now apparent with the Technical Secretariat subordinate to the Executive Council which is in turn representative of the Consultative Committee. Since 1983 much of the debate has focused upon the duties of these bodies, their relationship to each other and, in the case of the Executive Council, its composition, and rules of procedure. In 1984, Ambassador Rolf Ekeus, presented the report of the newly named **ad hoc** Committee on Chemical Weapons as CD/539, in which Annex I presented the text of the proposed Chemical Weapons Convention as a series of articles in what has come to be called "rolling text" i.e. ongoing negotiations are expected to add to the text and allow for changes in existing text, such as the removal of brackets. The current rolling text is CD/782 and the functions of the Consultative Committee and related organs are found in Article VIII. This article, however, does not stand alone, and the interactions between it and the others must be closely examined as one attempts to develop a global machinery which will implement such a Chemical Weapons Convention.

The Current Draft Convention: CD/782

This is not the place for a detailed review of all the articles of the "rolling text" but, in the context of the implementation machinery of a Chemical Weapons Convention, it is useful to indicate the verification provisions which will generate the inspection requirements of a Technical Secretariat. Articles III, IV, V, VI and IX all make specific demands of the Consultative Committee and its suborgans. Table 1 rearranges these requirements slightly to make the tasks clearer.

The following, then, are all verification tasks mandated by the Chemical Weapons Convention, and the machinery to implement them has to be developed out of Article VIII which outlines the current thinking with respect to the Consultative Committee, Executive Council and the Technical Secretariat. Articles which have not yet been considered in detail, i.e. Articles X (Assistance), and XI (Economic and Technological Development),

TABLE 1
CD/782 - Functions of Consultative Committee

	Programme	Verification
1.	Chemical Weapons Articles III and IV	Declarations Storage Destruction Transfer
2.	Chemical Weapons Production Facilities Articles III and V	Declarations Cessation Closure Destruction/Dismantling Temporary Conversion Reconstruction Transfers
3.	Activities not Prohibited Articles III and VI	Declarations Research and Development Permitted Production Non Production Transfers
4.	Challenge Inspection Article IX	CW Storage CW Production Allegations of Use
5.	Consultations Article IX	Compliance Inquiries

will undoubtedly add to the burdens already assigned under Article VIII. There are also many associated activities not specifically mandated by the above articles, e.g. compiling and updating lists of chemicals. The form of the international organization will have to follow the functions expected of it and it must be able to function as soon as the Convention enters into force. Chart 1 shows some of the interactions envisaged between the articles of the current draft Convention.

Organizational Structure for Implementation

It should now be clear what we expect our institutional machinery to accomplish. It should be in place by the time the Convention enters into force and have an inspectorate available to monitor its implementation and measure the continuing compliance of States Parties to the Convention. It must be able to deal with all of the issues outlined in Table 1. All this on a global scale of as-yet-undetermined magnitude, with the added complication that the CWC is "front-end loaded" since the destruction of chemical weapons and production facilities should be completed within the first ten years

In the period 1980-1984, the idea of an international inspectorate responsible for verification and under the control of an Executive Council became firmly entrenched as the only way in which all States Parties could be involved in "effective measures for international control and supervision" of a Chemical Weapons Convention. The first attempt to work out how it could be done is due to the Netherlands and is described in CD/445 of

CHART 1

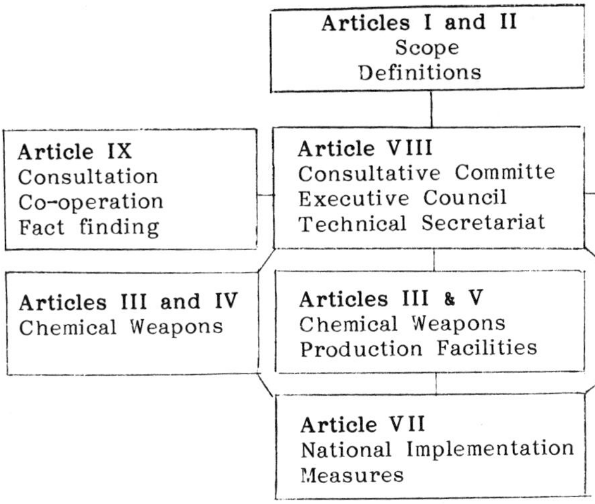

Articles I and II
Scope
Definitions

Article IX
Consultation
Co-operation
Fact finding

Article VIII
Consultative Committe
Executive Council
Technical Secretariat

Articles X & XI
Assistance
Development

Articles III and IV
Chemical Weapons

Articles III & V
Chemical Weapons
Production Facilities

Articles II & VI
Activities Not
Prohibited

Article VII
National Implementation
Measures

1984. The paper pointed out the need to reflect on the structure and size of an inspectorate which was expected to verify compliance and the uncertainties in such estimates due to a lack of knowledge of the final verification provisions, the numbers of chemical weapons and plants to be destroyed, and the level of inspections required by the chemical industry. Their conservative assumptions lead to a requirement for 50 permanent inspectors with 90 support staff and an additional requirement over the first 10 years for a further complement of some 75 to 115 inspectors and a 100 support staff. A UK paper was presented in 1985 and it dealt with the problems of nonproduction, making use of the safeguard experience of the IAEA. Thus, CD/575 further developed the ideas of the Netherlands paper with regard to routine inspections. In a further important paper, CD/589, the UK delegation discussed the problems of the constitution of an International Organization and paid considerable attention to the role of a Director-General in such a structure. They also proposed the establishment of a special pool of inspectors to undertake Challenge-Inspection. In the 1987 summer session, the UK presented CD/769 in which they articulated their concerns about the effective implementation of a Chemical Weapons Convention. They argued that there is a need for an organization to be operating immediately the Convention comes into force and so there must be a mechanism for putting a substantial part of a verification organization into place before that date, if the Consultative Committee is to meet its objectives.

The Machinery

Chart 2 shows the relative positions of the organs required to implement the provisions of a Chemical Weapons Convention.

The Consultative Committee is established as the main organizational device of the Chemical Weapons Convention. The only major item undecided is the decision-making process - questions of procedure, questions of substance, decisions taken by consensus or by simple majority or by two-thirds majority.

CHART 2

CONSULTATIVE COMMITTEE
|
EXECUTIVE COUNCIL
|
TECHNICAL SECRETARIAT
OR
INTERNATIONAL VERIFICATION ORGANIZATION
/ | \
HEADQUARTERS STAFF INSPECTORATE TECHNICAL SUPPORT STAFF

The ways in which it will interact with the other bodies is not altogether clear in the current text. The text also suggests that the Executive Council and the Technical Secretariat will be activated after the first meeting of the Consultative Committee but it is essential that this be changed to reflect the realities of the situation at that time. A recent suggestion is that the Consultative Committee be renamed the "General Conference". The Executive Council will depend upon agreement upon the composition and decision-making procedures of this body, as well as the extent of its direct involvement in some compliance issues. Most of the powers and functions of the Consultative Committee will be delegated to the Executive Council. There is no doubt that the Executive Council will be the central management authority with respect to the Chemical Weapons Convention. The Convention cannot function if this body is not effective.

There is agreement on the need for a Technical Secretariat which will directly manage the Inspectorate, but there is caution in the approach of delegations because of the uncertainties involved in its size, structure and cost. As can be seen from Articles III, IV, V, VI and IX and their respective annexes, the **ad hoc** Committee on Chemical Weapons has clearly spelt out what they feel has to be done by States Parties in order to remove the threat of use of chemical weapons. Confidence that these measures are being carried out lies at the heart of the verification regime. Within 30 days of the Convention entering into force, the Consultative Committee should have received declarations relating to Chemical Weapons, Chemical Weapons Production Facilities, Facilities related to Non Production, and these will have to be verified almost immediately. In the current Rolling Text the election of the "head of the Secretariat" and "appoint inspectors --" is in bracketed language under the functions of the Consultative Committee, and there is corresponding bracketed language under the Executive Council for "-- the head of the Inspectorate". We seem, therefore, to have the strange situation where the body of the text has not yet entirely sorted out the structure while the Annexes etc. lay out many duties which an Inspectorate will be expected to carry out! In a brief section on the Preparatory Commission it is stated that it would be expected to "appoint an Executive Secretary and establish a provisional technical secretariat --". It is clear that the **ad hoc** Committee is becoming increasingly aware of these problems and we can expect significant efforts to further elaborate text in the near future as evidenced by the commentaries in Appendix II of CD/782.

Institutional Machinery - The Next Steps

The degree of confidence in a Chemical Weapons Convention will in many ways be a function of how well the verification machinery monitors compliance. This means that an International Verification Organization must be alive and well when the Convention enters into force. The only way in which this can be done is if the **ad hoc** Committee, in its further negotiations, makes the elaboration of Article VIII a priority and gives the organization which results the necessary scope to verify Articles III, IV, V and VI. It must decide those questions of principle which belong in the Convention, set up the appropriate annexes and delineate clearly these tasks which must be carried out by a Preparatory Commission. The time has also passed when answers which are purely qualitative can be given, e.g. it is possible by task analysis etc. to establish the kinds of inspectors required but the numbers will be a function of the number of tasks to be carried out. In the absence of complete information on Stockpile and Production Facilities, it is impossible to develop a better than ballpark number for the inspection requirements. In CD/41 the Netherlands delegation developed a questionnaire which greatly assisted the organization of a draft convention, in that the answers provided by many delegations helped structure the thinking with respect to these elements required in the scope and verification of a Convention. A similar initiative in the area quantitative information is needed at this time.

The following questions must be answered soon:

(i) Size and structure of the International Organization
(ii) The number of inspectors required for initial inspections
(iii) The equipment needs of the organization
(iv) Cost estimates for the organization
(v) Training of Inspectorate

Each of the above will have to be discussed in depth with the appropriate quantitative background, but the following notes might help in the discussion.

(i) In 1985, the IAEA Safeguards Division had 258 staff members with 166 support staff. There were 180 inspectors who carried out 1980 inspections of 514 facilities and this took 7750 man-days, i.e. each inspector managed 43 inspection days.

(ii) The initial number of inspectors will relate to the number of States Parties required for entry into force since this will mandate the initial number of declarations etc.

(iii) This requires a great deal of specialist study. There is a good deal of data available in the Finnish Blue Books, the Norwegian Reports and the recent Finnish Workshop. These could form the basis of an integrated study which must be carried out as soon as practicable.

(iv) The IAEA corresponding budget was $33 million 1985.

(v) The training and recruitment of the inspectorate must be the subject of detailed study. A detailed programme must be in place before entry into force and will require the cooperation of industrial states and the possessors of chemical weapons, or related technology.

Conclusions

The work on developing the machinery necessary to implement a global Chemical Weapons Convention is well underway. The tasks of an International Inspectorate have been defined in the appropriate articles and the nature of two of the organs, the Consultative Committee and the Executive Council fairly clearly delineated. Further work on the size and structure of the Inspectorate requires the potential States Parties, and certainly those involved at the Conference on Disarmament, to be more forthcoming on the extent of chemical weapons stockpiles, their production facilities, if any, and the potential for the production of scheduled chemicals. It would also be helpful if delegations to the CD were to disclose how they plan to deal with Article VII, and the relationship they envisage between their National Authority and the Technical Secretariat.

Appendix: The Technical Secretariat (IVO)

Table 1 lists all of the verification activities which the Consultative Committee (General Conference) will be responsible for overseeing. Clearly this can only be managed by a subsidiary body which has a great deal of technical and scientific expertise. The actual size of such a body can only be estimated because insufficient data are available to quantify the needs. What can be done is a detailed analysis of the skills, personnel and resources that a Technical Secretariat will require. The basis for such an analysis is the inspection requirements of Articles III, IV, V, VI and IX; only the first four are currently amenable to in-depth analysis because of the uncertainties associated with IX. A crucial point is the kind of verification schemes which can be adumbrated but analysis does show that a variety of functions are common to such schemes and further that specific functions can be correlated with the skills required to carry them out. Hence, although the activities of the inspectorate may differ from programme to programme the number of viable inspectorate activities is finite. Once one has developed a coherent set of verification methodologies, a series of restrictions has to be applied; some of these will be physical, e.g. a requirement for tamper-proof equipment, but many will be political, e.g. degree of intrusiveness of verification.

The derived skills from such analysis show a need for engineers - chemical, industrial, computer and process - chemists, toxicologists, industrial hygienists, material accountants. There will be a need for a technical support staff including interpreters, computer and data communication technicians, electronic and instrumentation specialists and laboratory lawyers, accountants, and secretaries. Another group will have to be responsible for the training of all kinds of personnel. The actual size of the inspectorate will, in the first instance, be a function of the number of facilities to be inspected, the geographical spread of these facilities and the requirements of random inspection. The size of the support staff will be a function of the form of data management, the instrumentation employed, and the nature of the laboratory facilities utilized.

The costs associated with a Technical Secretariat will be a function of all the above, and will also include the administrative requirements, the location of the Headquarters and the travel costs of the inspection activities.

Adequacy Versus Feasibility in the Scope of the Projected Chemical Weapons Convention

Julian Perry Robinson

Introduction

The treaty to ban chemical-warfare weapons, namely the projected Chemical Weapons Convention being sought in Geneva at the 40-nation Conference on Disarmament, is now in the decisive phase of negotiation. An important consequence is that interest-groups liable to be disaffected by the treaty are becoming increasingly active in a number of countries. Public expressions of outright opposition to the treaty are still rare, but more subtle forms of antagonism are evident. One of their effects is to impose continual pressure on the scope of the projected treaty during its development by the negotiators, squeezing or otherwise distorting the comprehensiveness which the participating governments have declared as their aim. Friends of the Chemical Weapons Convention therefore need to muster safeguards of some sort in order to ensure that the treaty does not thereby become emasculated - recognizing, however, that the support of at least some affected interests is necessary for the successful conclusion of the treaty.

This paper addresses then the tension between desirability and feasibility: the constraints which expediency always tends to impose upon normative action.

The Avowed Purpose of the Treaty

Success in the present negotiation will take the form of consensus on the form of a new treaty regime which the negotiators think they would rather be inside than outside: a new state of international cooperation preferable, in the judgement of the participating governments, to the **status quo**. Both the vehicle and the characteristic of this preferred state is to be the elimination of chemical weapons. Like-with-like chemical-warfare deterrence, being the chief feature of the **status quo** that would be abandoned, is to be traded for a regime of chemical warfare disarmament. Through negotiations on scope and, especially, verification, the perceived reliability of the regime is to be built up to the point at which the competing option of chemical-warfare armament would seem to afford a lesser degree of security. The trade would be worthwhile.

The Preamble of the Treaty currently envisaged in the rolling text does not express the goal in such relative or political terms. What it says is this:

> ... **Determined** for the sake of all mankind to completely exclude the possibility of the use of chemical weapons through the implementation of the provisions of this Convention, thereby complementing the obligations assumed under the Geneva Protocol of June 1925...

It is from this language - this **normative** language - that the scope of the treaty is to be seen to have been derived. The primary determinant of scope is the manner in which those "chemical weapons" are to be defined. The secondary determinant is the extent of the "provisions" whose implementation would achieve the postulated objective, that of "completely excluding" the possibility of use. These two determinants are now discussed in turn.

(a) What are the 'chemical weapons' which the treaty should ban?

What **might** be regarded as a 'chemical weapon'? The Geneva Protocol uses the words "asphyxiating, poisonous or other" to describe the weapons whose use it proscribes. In the equally authentic French text, the words are "asphyxiants, toxiques ou similaires". If the Convention is not to impugn the Protocol, its scope must be no narrower. Chemical weapons must, at least, be toxic weapons.

But might the scope not be broader? During League of Nations days, incendiary and flame weapons were often subsumed with toxic weapons into a general category of 'chemical weapon', and even today newspaper reports of napalm use are sometimes headlined 'chemical warfare'. Further, if the eminent jurist Georg Schwarzenberger could become persuaded that the use of nuclear weapons would, by virtue of the radiotoxic fallout, violate the Geneva Protocol[1], should not these weapons, too, be subsumed within the category? And maybe conventional explosive weapons as well, since high explosives are chemicals?

If, however, we stay with the Protocal and draw the line around weapons whose primary mechanism is toxicity, we must remember that there are many different ways in which toxicity can manifest itself or, indeed, be defined. Not all are ways that necessarily correspond to a definition of 'chemical weapons' that is proper for arms-control or treaty purposes. For example, the 'acute' toxicity of a nerve gas may be differentiated from the 'chronic' toxicity of a carcinogen. Again, toxicity may threaten vital function or its threat may be limited to cognitive function, say, or motor function: it may, in other words, be 'lethal' or 'non-lethal'. Likewise, its effects may be transient or permanent: a tear gas is no less of a 'toxic agent' than a nerve gas. And should we be thinking only about toxicity towards human life, or should we also admit plant life into consideration, just as the conferees did in Geneva in June 1925? That would put chemical herbicides within the definition. There is scope enough, and perhaps need enough, for drawing lines around differentiated forms of toxicity. The increasingly derided use in Geneva of 'toxicity criteria' is one illustration of what can be done. Such criteria need not be technical ones, of course; military or political criteria may also serve. What, for example, of the 'toxins' that are already covered by the 1972 Biological Weapons Convention - not forgetting that certain toxins have been used in tear-gas weapons?

We do not have to leave the task of deciding all these points to the interplay of competing interests. For some of them, desirability clearly says no; for others, yes. Tear gases, herbicides and maybe toxins are the only really contentious matters.

Yet to be explored, however, is the question of **adjuvants** - those 'chemicals intended to enhance the effect of the use of such (i.e. chemical) weapons' at present contemplated for express prohibition in a square-bracketted part of Article II of the draft Convention. An example of such an adjuvant, and once nominated as such in the Annex to Article IV, is that of the thickeners used to modify the dissemination characteristics of liquid agents, such as sarin in the US arsenal, and lewisite, soman and VX in the Soviet arsenal. Stabilizers and penetrants would be other examples.

The last time the international community wound itself up into any form of judgement on the definition issue was in 1969, on the occasion of the 'Swedish' resolution in the General Assembly. By resolution 2603A (XXIV) the Assembly defined "chemical agents of warfare" as "chemical substances... which might be employed because of their direct toxic effects on man, animals or plants". The vote was 80 to 3 with 36 abstentions. Although most of the abstentions (which included all of NATO except for Portugal and the United States, which voted against, alongside Australia) were explained on procedural rather than substantive grounds, this was hardly a ringing affirmation of the so-called 'extensive' definition. Even so, the definition is probably one with which the great majority of governments would concur today.

(b) What is the relevant historical experience?

One may estimate - hesitantly - from the available sources that the quantity of chemical-warfare agents (in the sense of the Swedish resolution) that have been used in war this century is somewhere between one-fifth and one-quarter of a million tonnes.

Nearly half of that quantity has comprised herbicides targetted on food crops, economic crops or natural cover. This is a chemical-warfare technique that was first studied in earnest during World War II, and first practised, in Malaya and then North Africa, during the 1950s. It was brought to an astonishing pitch of destructiveness during the Vietnam War.

Maybe one-tenth of the century's tonnage has comprised tear gases and other sensory irritants, beginning with the 15-30 different ones used in World War I from its opening month onwards. An official history of the war records that there were occasions when soldiers shot themselves in order to escape from the pain of the 'direct toxic effects' of some of these chemicals. Rarely employed in subsequent wars, several thousand tonnes were nonetheless used in the Vietnam War.

The remainder of the century's tonnage has comprised the 'poison gases' of popular parlance, the antipersonnel casualty-causing chemicals whose initial use quickly followed that of the irritants during World War I. Some 15-20 different ones were tried out then. At least a million people fell victim, most of them either to the asphyxiants chlorine, phosgene and diphosgene or, during the final 16 months of the war, to the vesicant known as 'mustard gas'. It is mustard gas that has caused many thousands of Iranian and, more recently, Iraqi Kurdish, chemical-warfare casualties in the Gulf War.

Untried (for the most part) in the arsenals exist the enormously deadly nerve gases, which are the successors to phosgene and, in many but not all of its roles, mustard gas. There also exist, but apparently more in concept than in actuality, the incapacitant casualty agents: supertoxic chemicals whose effects disable rather than kill, and from which the incapacitation may be transient, though lasting much longer than that from the irritants. Psychedelics, paralysants, anaesthetics and several other types of drug have been studied in this regard, and sometime even stockpiled as weapons. Development programmes in all these and other areas continue. Recent commentaries warn against the possibility of nasty surprises, certain of them even suggesting that dramatically new forms of chemical weaponry are already emerging in secret military programmes[2].

The Current Situation on Scope in Geneva

So, in the face of this diversity, what have the negotiators actually been doing in defining 'chemical weapons'? The present rolling text[4] gives some indication, though as yet little clarity. (It is nearly time, however, for Article II to come round again.)

One may look first at Geneva's draft control lists of particular chemicals that are scheduled for special attention. There are 40-odd chemical families in these schedules. None of them are herbicides. Two are supertoxic riot-control agents, but they are listed, not in the schedules proper, but in a 'to be discussed further' addendum. The same is true for the one listed toxin. So we seem to see here forces of exclusion, not inclusion, at work.

But the schedules are draft only, and in any case relate to the verification provisions. They are not intended to limit the scope of the general proscriptions to be laid down by the treaty. Here, the negotiators avow that the so-called 'general purpose criterion' is paramount. It is contained in Article II, where at present it is, however, much beset by qualifications in footnotes and bracketted alternative language. Even so, it gives the draft treaty an extremely broad scope. It **nearly** says that the treaty is to ban possession of all chemicals unless they are intended for purposes not prohibited by the treaty (which purposes the treaty is to define **explicitly**) and are of types and in quantities consistent with such purposes. "Nearly", because what the draft article actually speaks about is 'harmful' chemicals plus all chemicals that can take part in the production of such chemicals. That is really pretty close to saying 'all'.

There are those who will say that general statements, general criteria and general precepts in a treaty are all very well, but what really matters are the specific control lists which give the treaty teeth. Maybe. In which case it is most significant that, of those 40-odd entries in the draft control schedules, less than half are chemicals which the experience of history has shown to be chemical-warfare agents. The majority of them are 'chemical weapons' in name only - according to the very special definition now used in the draft treaty. Thus, that familiar school-laboratory reagent, phosphorus trichloride, is accorded the same status of 'chemical weapon' as phosgene, being listed in the same control schedule. This is because of its utility as a production intermediate. And notwithstanding its use in school laboratories it is to be subject to Schedule (3) verification. Is this, one may ask, an instance where considerations of what is desirable in the treaty have run way ahead of what is feasible? Or, conversely, does it simply mean that the Schedule (3) regime is mere facade: that the real killers which it contains will not in fact be significantly constrained - recognition, in other words, that what is desirable may have to be suppressed in the interests of what is feasible?

Impingement upon Civil Chemical Industry

No doubt one must look to the American binary-munitions programme for the originating impetus behind that transformation of production intermediates into 'chemical weapons'. If all it takes is a built-in canister of powdered sulphur to convert a bomb load of ethyl diisopropylaminoethyl methylphosphonite into a devastating nerve-gas weapon, then the treaty should obviously place very stringent controls upon the otherwise rather harmless chemical, and maybe even on sulphur as well. Washington has hardly been in a position to oppose the presence of such precursors in the control lists even if it wanted to, and there was of course no shortage of people insisting upon inclusion. Yet the feasibility of binary technology had in fact been established, and recognized by CW specialists outside the United States, well before the American binary programme had come anywhere near its present mass-production phase - meaning that the chemical treaty would anyway have had to accommodate it. And the present Gulf War provides a lesson no less clear-cut on the importance of controlling production intermediates. If there is a case for having the O-ethyl 2-m,n, diispropyl amino ethyl methyl phosphonite (QL) just mentioned in the treaty control lists, Iraq has been showing us very clearly why the thiodiglycol which it has been importing - not a binary reactant, but nonetheless one simple step short of mustard gas - should be there too.

There is one important difference between QL and thiodiglycol: the latter has significant application in civil manufacture and commerce. So do several of the other precursors that have thus far been scheduled in the rolling text. Civil chemical industry has thus been brought squarely into the problematique of chemical-warfare arms control. It has, since 1985, been assuming the role that had long been foreseen for it, that of principal actor in the negotiations.

An effect of this has, of course, been to mobilize a further set of interests that must somehow be accommodated if agreement is to be reached. One might say, then, that progress towards the Chemical Weapons Convention has now become overtly hostage not only to the various interests that are active in the defence and foreign-policy fields but also to those industrial and commercial interests that, quite understandably, feel threatened by the demands of regulation, data-disclosure and inspection which the treaty's verification regime may impose.

Trial runs within the Dutch and Australian chemical industries suggest that in fact the risks to, for example, the competitiveness of individual corporations can be held to tolerable levels. But doing so is bound to place scope under continual pressure: such difficulties as the proper alignment of data-reporting thresholds to adequate specifications of militarily significant quantity, or what is and what is not to be treated as a 'chemical weapons production facility', these and other such difficulties will most probably be resolved in favour of looser rather than tighter constraints on industry.

Observe, for instance, what has happened to that largely hypothetical 'Schedule (4)' during the interval between the January 1988 and the April 1988 rolling texts[4]. In the former it had been envisaged that additonal categories of civil-use chemicals might need to be brought under control over and above those already treated in Schedules (1), (2) and (3): hence the fourth annex to the article of the draft Convention dealing with civilian chemical industry (article VI), an annex then entitled 'Commercial production of toxic chemicals, not listed in Schedules (1), (2) or (3) that might be relevant to the Convention'. Come April, the scope of the annex had become diminished to 'Production of super-toxic lethal chemicals not listed in Schedule (1)'. Should one take it, therefore, that there are no products of, for example, the veterinary-drugs industry which threaten to become supertoxic chemical weapons of the incapacitating, as opposed to lethal, variety? Or that commercial availability might not itself become a significant incentive to adapt a less-than-supertoxic lethal chemical into a chemical weapon. Is this, in short, the sort of price that has to be paid for the industry's support of the negotiations? Very probably, as it happens, not; but, if it were, it would imply that the industry stood to gain nothing at all to its own benefit from the Chemical Weapons Convention: That Halabja, for example, was an episode utterly different from Minimata, Seveso or Bhopal.

Secondary Determinants of Scope

Constraining possession of chemicals that could serve as, say, nerve-gas precursors, looks like a necessary device for achieving implementation of the norm. In other words, without controls on civil chemical industry, there might well be insufficient confidence in the treaty regime. One can think of several other activities that might, for the same reason, be brought within the scope of the treaty: the development of chemical weapons, most obviously, should be banned expressly; also their transfer.

These are positive constraints, so to speak. It may also be necessary to envisage negative ones, which is to say express exemptions from the scope of the treaty's prohibitions. Antichemical protective measures are the main case in point. The argument for leaving them outside the treaty altogether is surely unassailable, however much their inclusion might ease the verification problem. For not to exempt them would be to require states to maintain themselves in a perpetual state of vulnerability towards potential cheaters. The overall regime would thus be most unstable, and confidence in it liable to be evanescent. Maybe article X will turn out to offer an unexpected safeguard in this respect.

Such exemptions - and others, too, can be envisaged, as France has demonstrated - of course provide a further way of accommodating recalcitrant interests. They are thus yet another threat to extensiveness of scope. Since 1978, or thereabouts, governments participating in the chemical-weapons talks have been in agreement that the projected treaty should be 'comprehensive'. That, very possibly, is how it will indeed look, thanks to the general purpose criterion. But as regards actual teeth - those parts of the treaty's scope with which compliance is actually monitored - there are bound to be very many gaps. The appearance will be comprehensive, but the reality will be partial. The scope of the verification system is liable to be very much narrower than the scope of the treaty itself.

The is not necessarily a bad thing. Of course it will be impossible to monitor every chemical that could be made into a weapon, or every factory where such a chemical could be produced, or every activity that might contribute to offensive chemical-warfare capability. But the need is not one of creating a watertight and inviolable regime that will make chemical warfare impossible. The need is rather for a system allowing states-parties to test one another's continuing good faith in participating in the treaty regime, and in subscribing to its underlying norm. Above all, the need is for a system that will diminish those areas of ambiguity which 'national technical means' and such like might otherwise misperceive as violation serious enough to endanger national security - particularly ambiguities residing in the 'dual-purpose' chemicals and facilities that occur in civil industry. Quite short control lists may be perfectly adequate for so modest an objective, provided the overall system included an international inspectorate of proper independence, powers, integrity and skills. A sufficient non-production verification system might thus be one that made no pretence at covering all evasion possibilities but which spread out the sand, so to say, upon which treaty-noncompliance would probably leave some sort of footprint, were it to be happening. Feasibility and desirablity may well be in harmony.

However, the regime must, in one important sense, be a living organism. There will be developments, as time goes by, in the military, political and technological worlds within which the regime is to function, and these may be stimuli to which it must be capable of responding. As those recent commentaries noted earlier show, it is not impossible to envisage developments that might make chemical armament considerably more threatening than it is today. In that event a strengthening of the regime could become essential if the whole structure were not to collapse. It would be disastrous, in such contingencies, to discover that the general purpose criterion had become so attenuated by default or by concessions to interest groups that it was no longer an accepted marker of scope - a sufficient mandate for the International Authority to augment the control lists or do whatever else was necessary.

The Primary Safeguard?

A ready mechanism for revising control lists and amending control procedures is clearly a primary requirement. The role here of the projected Scientific Advisory Board or Scientific Council will be of central importance. But how is such a body to be sufficiently isolated from political pressures to inspire confidence, yet at the same time sufficiently influential to ensure action? And what should be its precursor at the crucial stage of the Preparatory Commission? These are all questions which people in Pugwash may be especially competent to address.

References

1. G. Schwarzenberger, The Legality of Nuclear Weapons, London, 1958.

2. See, especially, J. Hemsley, The Soviet Biochemical Threat to NATO: The Neglected Issue, London: Macmillan, 1987.

3. The latest version of the 'rolling text' (the most recent edition of the draft Chemical Weapons Convention in which the CD Ad Hoc Committee on Chemical Weapons registers consensus and dissensus) is Appendix I of CD/831, 20 April 1988.

4. CD/795* of 2 February 1988, at page 80, as compared with CD/831 of 20 April 1988, at page 86.

Chemical Disarmament: Reliable and Efficient Control

Kirill Babievsky and Nikolai Enikolopov

Introduction

Banning chemical weapons is central to stopping the arms race, reducing the danger of a new war, and attaining disarmament. Together with nuclear and bacteriological weapons, chemical arms are the most hideous instruments of mass murder. Being relatively simple to produce, these weapons are easily accessible to any country. Meanwhile, given the technological level and scale of their development, chemical weapons can inflict losses far surpassing the human toll in any feasible conventional conflict. Hence, the immediate removal of the chemical war threat is a task of paramount importance, an effort that requires the joint endeavours of all nations.

Today, after nearly two decades of negotiations, the preparation of a universal International Convention which would place a general and total ban on chemical weapons, eliminate CW stockpiles and the industrial potential for their production, is nearly complete. In June 1988, the Third Special Session of the UN General Assembly on Disarmament had the possibility to examine the draft convention - a thick pile of documents reflecting both agreed matters and unsolved problems. Several rounds of Soviet-American consultations were held parallel to the multilateral negotiations for the elimination of chemical weapons. The USSR and USA worked out a draft set of provisions concerning the elimination of facilities for the production of chemical weapons, and prepared documents on additional bilateral measures to enforce the convention: a Soviet-American consultative mechanism; and participation, on a mutual basis, of Soviet and American representatives in international tours of inspection in the territories of the USSR and USA, which are specified by the convention.

Control Issues

Proceeding from its basic conviction that reliable and efficient control over real disarmament is necessary, the Soviet Union announced, in August 1987, that its delegation will seek legal endorsement of the principle of obligatory inspections on request, without the right to deny such inspections. This approach takes us much closer to the solution of the pressing political problem facing the chemical weapons talks: on-site inspections on request in case non-compliance is suspected.

However, control and verification are not an end in themselves, they are only a measure guaranteeing safety of participating sides, and protecting the interests of all parties to the Convention. Disarmament without verification is impossible, just as verification without disarmament is devoid of sense. This points to the conclusion that verification methods, and above all, the scale and scope of control, should be 100 per cent adequate for the practical steps undertaken to curb chemical weapons.

It is our belief that control with national technical means retains its crucial role. Experience of the use of national means to check compliance with earlier treaties confirms their obvious reliability and indisputable authority. In addition, given the technological development of national verification means over the last few years, the system of control with national technical means will remain, for many years to come, the central element of control over compliance with international agreements.

At the same time, control and verification with national technical means should, in each individual case, be complemented international procedures to improve the reliability and efficiency of verification. We shall cite several examples of such international procedures.

a) <u>Verification of destruction of chemical stockpiles</u>

Here, control should be at its strictest and most objective. Inter national inspectors must, in the most thorough way imaginable, make certain that everything earmarked and brought to the designated site for destruction is indeed irreversibly eliminated. Hence, it is proposed to have several international inspectors at destruction sites constantly throughout the period of stockpile elimination. It appears that all parties are in agreement on this matter.

b) <u>Verification of allowed production</u>

Verification of allowed production of chemical agents, otherwise banned by the Convention yet required to test anti-chemical defence, such as medicinal and pharmaceutical preparations, must also be strict, perpetual and efficient. It is therefore proposed that the production of such chemical agents should be organized at a special facility limited to, say, an annual capacity of one tonne.

International inspectors are to remain at such facilities at all times to exercise strict control over the production process and subsequent traffic. This approach satisfies most parties to the negotiations, but not all. For instance, members of the US delegation proposed to allow all producers in need of small amounts of such chemicals to manufacture them themselves. With such an approach, any laboratory, company or testing facility, which is in need of a small amount of banned chemical agents, is given the right to produce them independently, after notifying the proper international authority. But who is to guarantee that a modern chemical laboratory would not be used for production of ultratoxic substances for chemical warfare? What strictness can one talk about with such a system of control? This could, in fact, lead to uncontrolled - and legalized - production of dangerous chemical agents.

c) <u>Control over producers of chemical weapons</u>

These procedures remain to be fully agreed upon by all parties to the negotiations. How is it possible to supervise effectively the closure of a weapons-producing facility? How will it be possible to make sure that production of such weapons is not re-started? These questions are difficult, yet not impossible, to answer. The Soviet Union has always believed and proved that control over the closure of chemical-weapons-production facilities can fairly easily be carried out by national technical means. This approach must, of course, be supplemented by passing into law the proposal for obligatory inspections on request without the right to deny such inspections. This approach is the most logical from the point of view of efficient control, since it leaves no possible loopholes. The stands of the USSR and the USA on this matter have almost converged, although the United States is still in favour of making a certain number of privatelyowned enterprises exempt from control, including transnational corporations with enterprises in foreign territories. But the Pentagon's main contractors for binary weapons are not to be found among leading US chemical companies; they are all small companies and enterprises, which are difficult to control with international inspections.

A number of participating countries made proposals concerning inspections on request. For example, in 1986 Britain came up with the idea of alternative measures in case full access to the facility under suspicion could not be made available. The verdict on the adequacy of the proposed alternative measures must be left to the country which makes the request. India, expressing the views of numerous non-aligned countries, submitted its proposals concerning inspections on request, in which the leading role, as opposed to Britain's approach, is given to the executive country authorized to grant such inspections. The delegation of the People's Republic of China adopted a similar approach.

In 1984 the US delegation proposed to the Soviet Union to exchange, on a bilateral basis, information on their respective chemical arsenals and production facilities with verification via random inspections until such time as a convention could be signed. At present, we have no objections to this proposal. There remain disagreements, however, on the schedule of exchange and number of inspections.

d) **Verification of non-production of chemical weapons by civilian enterprises**

It is clearly not enough merely to eliminate chemical weapons. Guarantees are needed against a return to chemical-weapon production anywhere in the world. Here the main difficulties pertain to the attempts of some Western countries to weaken the strict measures of control over the chemical industry, proposed at the talks, as well as to technical difficulties in elaborating effective control procedures at industrial facilities.

Is there a way out? Ours is an optimistic approach. The Soviet Union proposes an international experiment to test the procedures, which have already been negotiated, of systematic international verification of non-production of chemical weapons. This experiment could be carried out at chemical facilities voluntarily made available by the state. It is our belief that if effective chemical disarmament is to be a success, a ban must be imposed on all activities pertaining to development and production of chemical weapons at enterprises and organizations - both state and privately-owned - within the authority or jurisdiction of the country proposing to sign the future convention. This would also cover transnationals and their subsidiaries or affiliates. For this purposes it is proposed that all chemicals produced by civilian industry should be divided into four categories in the future Convention.

Category One would consist of supertoxic lethal chemicals with a compound influence characteristic of chemical poisons and part of binary-type chemical weapons. Supertoxic lethal chemicals similar to those in Category One, but produced for medicinal, pharmaceutical and other allowed purposes, are to make up Category Two. Category Three is to include key precursors of supertoxic lethal chemicals, which are used for allowed purposes. Chemicals of dual purpose, which are produced in industrial quantities for peaceful purposes - even though they may be components of chemical weapons - make up Category Four.

Depending on the volume of production of the above-described chemicals, and their importance as potential chemical weapon-grade materials, various control requirements are imposed on these four categories. For instance, a country which signs the convention must ban internal production of Category One chemicals, allowing their production only on special facilities at one location in amounts not exceeding one tonne per year and under perpetual international control.

Category Two chemicals are to be produced at enterprises appointed in advance, in quantities agreed upon by parties to the convention. All information on production and utilization of these chemicals shall also be made available, with control over utilization executed by international on-site inspections.

Enterprises engaged in production of Category Three chemicals shall also be declared. Information concerning production and utilization of such chemicals shall be submitted to countries participating in the convention. Control over non-use of these chemicals for purposes banned by the convention is to be carried out by means of perpetual international control of large-scale industrial production, with international inspections on request in case of small-scale production.

An identical form of control is proposed for Category Four chemicals. Along with presentation of data on all facilities producing such chemicals - location, overall amount produced, distribution in production, and so on - to an international authority (consultative committee), international inspections on request are also provided for. Installations producing such chemicals are to be placed under perpetual international control, and opened for inspection on request.

Analysis of the proposed method of control over civilian production shows that cheating will be next to impossible. Non-compliance will be established at once by international authorities. This approach is realistic and constructive, even though it does require a streamlined network of international control.

It should be taken into account that the ban on the production of chemicals limited by the future Convention should also cover chemical laboratories at educational institutions, research establishments and testing sites owned by the state or private citizens in those countries where chemical weapons research was or has been carried out since 1 January 1946, and where an appropriate laboratory-production potential exists.

Recently, there have been many statements to the effect that the greatest obstacle on the way to a speedy elaboration of a convention on chemical weapons is posed by control and verification questions. In fact, there are numerous problems to solve if we want to come up with a convention that would entirely obliterate a whole class of weapons of mass annihilation, and would keep it from being reborn in the future. Provided that there is a real desire to agree on arms reduction measures, control has never been - and, indeed, can never be - an obstacle.

Role of Scientists

In conclusion, we offer a brief comment on the international cooperation of scientists in the sphere of chemical disarmament. We scientists are well aware of the importance of eliminating chemical weapons and preventing their production in the future. Joint elaboration of efficient technical means of verification, and the standardization of analytical methods, instruments and procedures of control, can become an importnat sphere of international scientific cooperation. This cooperation could make use of the experience of successful Soviet-American seismic experiments designed to control nuclear weapon tests. It is our belief that the Pugwash Movement has all the resources to take the initiative and take the first practical steps which would promote a speedy conclusion of a convention on chemical disarmament.

Part V

European Security

Towards Conventional Stability in Europe

Albrecht von Müller

Introduction

This is not a traditional paper. It is impossible to write a comprehensive introduction on the issue of conventional stability without drastically exceeding the format of background papers for Pugwash Conferences. On the other hand, due to the complexity which is characteristic of conventional stability, even an introduction would not be worthwhile if it did not at least give an idea of the various determining factors and their interdependencies. For this reason, I chose to present a structured overview of the core aspects and arguments, instead of writing a traditional paper. In the following I will touch upon eight topics crucial for the understanding of conventional stability:

- the political goals behind the striving for stabilization in the conventional realm;
- the two military effects of conventional stability: deterrence against intentional attack and structural provisions against inadvertent crisis escalation;
- the net-deterrence equation, or why detente can enhance deterrence;
- the dynamics of crisis escalation and how it can be interrupted;
- an operational definition of conventional stability;
- the need for a much better integration of arms control policies and force modernization;
- outlines of a stability-oriented modernization;
- outlines of a stability-oriented arms control regime.

In those cases, where the tables/charts are not self-explanatory, a short description provides the necessary explanations.

The Political Goals Behind the Striving for Stabilization in the Conventional Realm

The first goal is to reduce the military threat potentials in Europe thus further reducing the probability of war by intention or crisis escalation. Secondly, there is the desire to provide a military regime in Europe which makes a return to confrontation-orientation politics by military threats virtually impossible, that is to create the military prerequisites for a far-reaching, irreversible process of détente and conflict transformation. A third goal is to free resources and reallocate them from defence spending to more constructive societal purposes.

The Two Military Effects of Conventional Stability: Deterrence Against Intentional Attack and Structural Provisions Against Inadvertent Crisis Escalation

For a long period NATO policy has been focused almost exclusively on deterring a Hitler-type attack by WTO. But there exists an alternative path to war, and that leads through crisis escalation. Unfortunately, much of what seems useful for deterrence (for example, strong counter-air capabilities), is at the same time highly disadvantageous regarding the dynamics of crisis escalation. Only a regime of conventional stability provides both: reliable deterrence and reliable provisions against crisis escalation.

The WTO defence posture of throwing the conflict immediately back on NATO territory is even worse. Because any alliance, after having received the initial blow, is still capable of throwing the conflict immediately back onto the aggressor's territory must obviously be offense-capable in the first place.

Therefore, it is essential to shift from a posture of deterrence through the threat of rapid escalation (NATO) and an offensive defence concept (WTO) to a conventional stability regime, in which neither side is capable of attack and invasion. This would be a big step forward.

The Net-Deterrence Equation

$$D_n = P_w + P_r$$

$$P_r > 0 \text{ increases } D_n; \; P_r < 0 \text{ decreases } D_n$$

(D_n = Net-Deterrence; P_w = prospects in case of war;
P_r = Prospects in case of renunciation of military force).

Even if we concern ourselves only with the deterrence effect as such, it is not enough to provide devastating prospects in case of war. What really counts is the difference between the future prospects in case of war compared to those if war is avoided. This means that threats to destabilize the opposing system anyhow by political or economic means directly undermines deterrence. On the other hand, prospects for a peaceful conflict transformation increase net-deterrence by adding to the difference between war and the renunciation of military force.

The Dynamics of Crisis Escalation and how it can be Interrupted (Fig. 1)

A crucial goal of conventional stability is to provide a hedge against crisis escalation. This requires that one has, at least, an idea of how a political crisis can get out of hand and escalate into war. The purpose of the extremely simple model here is to show just this.

Crisis escalation is described as two very simple optimization processes that are intertwined in a specific way. The dollar auction serves as a model, which makes transparent the characteristic intertwining of two decision-making processes and the crucial gap between "local" and "global rationality". In the dollar auction one dollar is sold. The specific rule is that not only the highest bid has to be paid, but also the second highest. This creates an incentive for the person who made the second highest offer to make a new, higher one, in order to reduce his losses. As the same logic applies to bidders, an interactional trap is constituted.

Dollar Auction Model

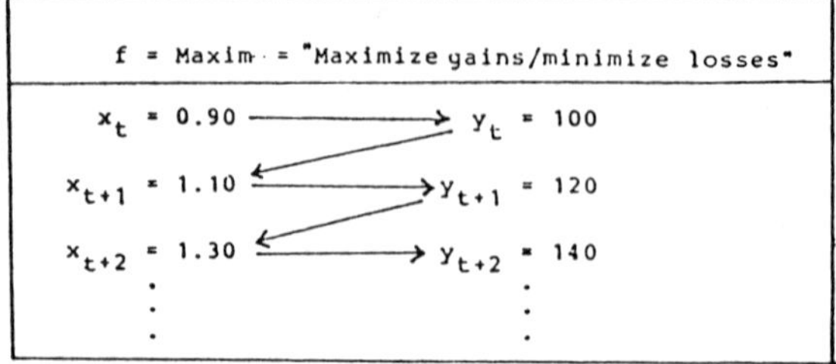

Conventional Crisis Scenario

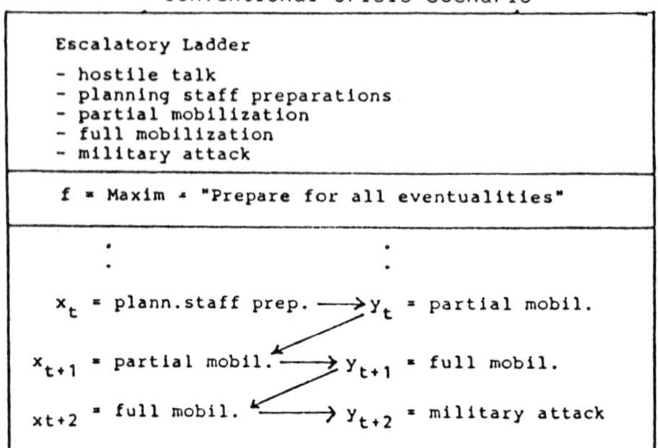

(The Dollar Auction has the specific rule that not only the highest offer has to be paid, but also the second highest. This creates an incentive for the one who had the second highest offer to make a new, higher one thus creating an interactional trap.)

Figure 1 The Dynamics of Crisis Escalation (I)

In a situation in which Y offers \$0.80, it is rational for X to offer \$0.90. If the auction would end at this point, Y would have to pay \$0.80 without getting anything and X would have to pay \$0.90, but would receive \$1.00, that is he would still have a profit of \$0.10. In this situation Y would behave rationally in offering \$1.00, thus having no profit, but at least avoiding a deficit of \$0.80.

At this point the critical threshold is passed. For X it is now rational to offer \$1.10, thus reducing his loss which would otherwise be \$0.90 to \$0.10. The same logic, of course, applies to Y. Instead of paying his last bid of \$1.00, and getting nothing, it is rational for him to offer \$1.10, thus reducing his loss to 20 cents.

This logic can theoretically be continued to infinity: comparing the end result at t, where the second highest offer was, it is always rational to get in front by offering \$0.20, thus reducing one's own loss by \$0.80.

In hundreds of empirical tests it took the players quite a while to understand this interactional trap, and then break out of it through compromise and by sharing the losses. In most cases, the dollar was in the end sold for a price between \$4 and \$5.

The argument here is that a similar logic applies to crisis escalation. In this case all sides also continue to take measures which "locally" seem to optimize their situation - but the sequence of actions of both players lead to a situation of escalatory dynamics, thus drastically worsening the overall situation. (To a certain extent the same logic also applies to the arms race as such. Here also exists the "fallacy of the last step", or, in other words, an interactional trap which is characterized by a gap between "local" and "global rationality".

An interesting point to be made in regard to crisis stability is that through such action sequences, it may be possible to destabilize a situation which at first glance seems to be extremely stable. In the example given in Fig. 2, we have a game theoretical pay-off pattern with an absolutely dominant advantage for cooperative behaviour (9:9) and a clear deterioration for both actors in case of conflict (1:2, respectively 5:1).

1) The Pay-off Patterns under a CS-regime

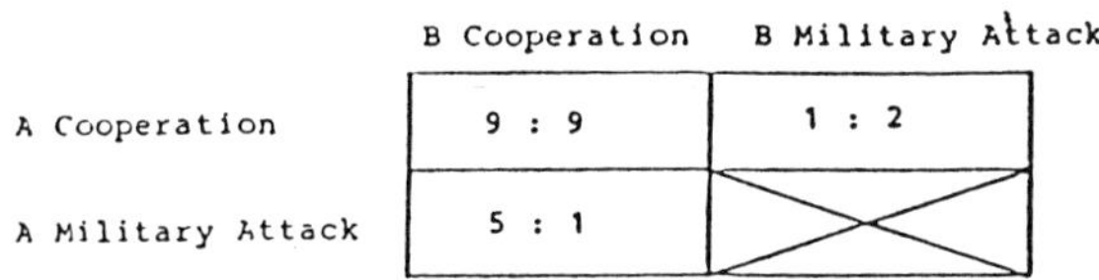

	B Cooperation	B Military Attack
A Cooperation	9 : 9	1 : 2
A Military Attack	5 : 1	X

2) The Critical Threshold of War Probability(p_w)

for A: $1(p_w) + 9(1-p_w) = 5; \quad p_w = 0{,}5$

for B $1(p_w) + 9(1-p_w) = 2; \quad p_w = 0{,}875$

3) How p_w spirals up

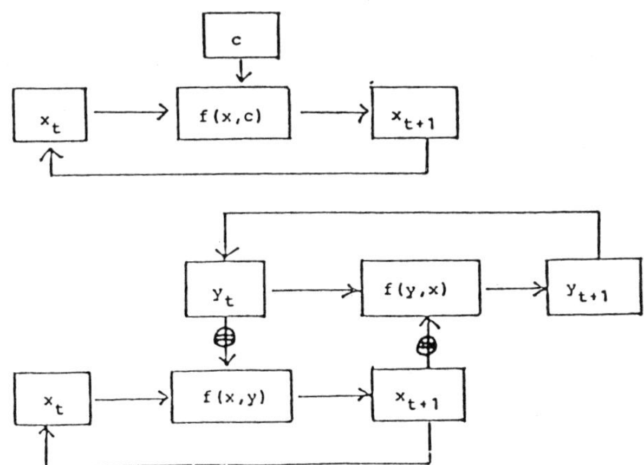

⊕ Possible interruption of escalatory loop by defense dominance

Figure 2 The Dynamics of Crisis Escalation (II)

This defines a fairly high threshold of war probability - and only beyond this point does it become rational to attack in order to gain at least the bonuses of pre-emption and attack (0.5, respectively 0.875).

Given such a clear advantage for cooperative behaviour one would normally see no reason why the critical threshold to war probability should be crossed. Yet, through the process of crisis escalation and mobilization dynamics, exactly this could happen - following the same logic that makes people buy one dollar for the price of several dollars. Analysing the structures of this interactional trap we find two simple, but intertwined iterative loops in which the result of one action phase becomes the input for the next.

The only way to break out of this vicious circle is to reduce the relevance of the transfer from the Y-loop to the X-loop and vice-versa. And this means making sure that mobilization processes are not increasing the threat for the opponent and that the bonuses for pre-emption and attack are systematically minimized.

An Operational Definition of Conventional Stability

Conventional stability is not provided just by "equilibrium", "balance", "symmetry" etc. Otherwise a duel situation would be a highly stable one - which it is obviously not, due to high bonuses for pre-emption and attack which exist despite the perfect symmetry. Conventional stability exists when, and only when, neither side can attack successfully. For such a regime we introduced the notion of "structural incapability to attack" (SIA).

Obviously, one can attack even with an old kitchen knife, given the opponent is completely unarmed. This demonstrates two important points: something like a purely defensive weapon simply does not exist; SIA is predicated on a military regime but never on the quality of one's side forces in isolation.

Despite the fact that all military forces have both a specific offensive capability and a specific defensive capability, the emphasis between these can be shifted quite substantially. Stability exists then, and only then, when the robust defence capabilities of both sides exceed the offence capabilities of the opponent. Therefore, we define conventional stability as a regime of mutual, robust defence dominance.

There are five aspects and criteria of conventional stability:

- "Pre-Emption Stability": the possible bonuses for pre-emption must be minimized. This can be achieved by three means: the reduction of the number and relevance of targets for pre-emption; improved protection for the remaining, inevitable targets; the limitation of weapon systems that can be used for pre-emptive strikes.

- "Battlefield Stability", part 1: the exploitation of the advantages of the aggressor on the battlefield must be inhibited. This means especially that surprise effects and local superiority must be denied. It can be achieved by limiting offence-prone mobility, massing and momentum.

- "Battlefield Stability", part 2: the structural advantages of the defender must be optimally used. This means that there should be no arms control limitations to non-mechanized infantry, terrain preparation, barriers and mine technology. The same holds true for anti-tanks and anti-air PGMs as long as they are not mounted on platforms.

- "Force Generation Stability": the potentials for force generation must be controlled and shaped in a way that no significant superiority can be achieved at any time or by any mobilization pattern.

- "Arms Race Stability": the overall military setting must make sure that the acquisition of additional offensive capabilities is far more expensive than the means to compensate for them.

The topology of "force-ratio development functions" as an indicator of crisis- and arms control stability (Fig. 3)

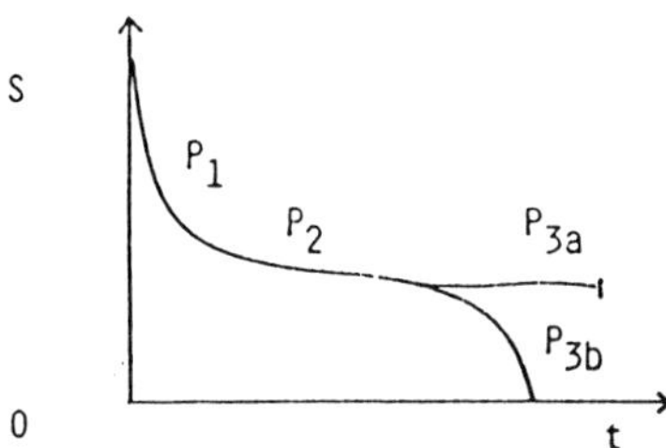

S: Strength of the defender
P1: Starting-phase with over-proportional losses of the defender due to effects of preemption
P2: Mid-phase of conflict with fully developed forces on both sides
P3a: End-phase with stabilization of defender and break-down of offense
P3b: End-phase with accelerated break-down of the defence
t: Time

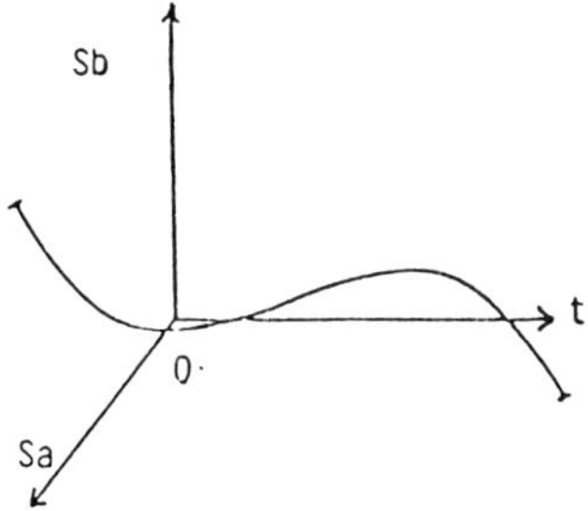

Sa: Strength of side A; Sb: Strength of side B.

Figure 3 The Topology of "Force-Ratio Development Functions" as an Indicator of Crisis- and Arms Control Stability

The purpose of these "force-ratio development functions" is to show the development of the force ratio over a certain time sequence during a conflict. If we attribute one axis to the aggregated military strength of A, one axis to the aggregated military strength of B, and one axis to the time factor, we get a curved trajectory in a three dimensional phase space. In order to simplify the graphical representation, only the axes for A and B are kept in all the following figures. (This can be interpreted as turning the three-dimensional coordinates system in such a way that the time axis points directly to the observer or away from him, thus seemingly disappearing.) In all the following cases we have two functions: one for the case if A would attack, and one for the case if B would attack. The shaping and positions of these functions indicate clearly the stability features of a given military regime. Only if the functions intersect before crossing the axis, are the initial advantages for the attacking side overcompensated, i.e., the defender wins the war. Four basic stability variants are shown in Fig. 4.

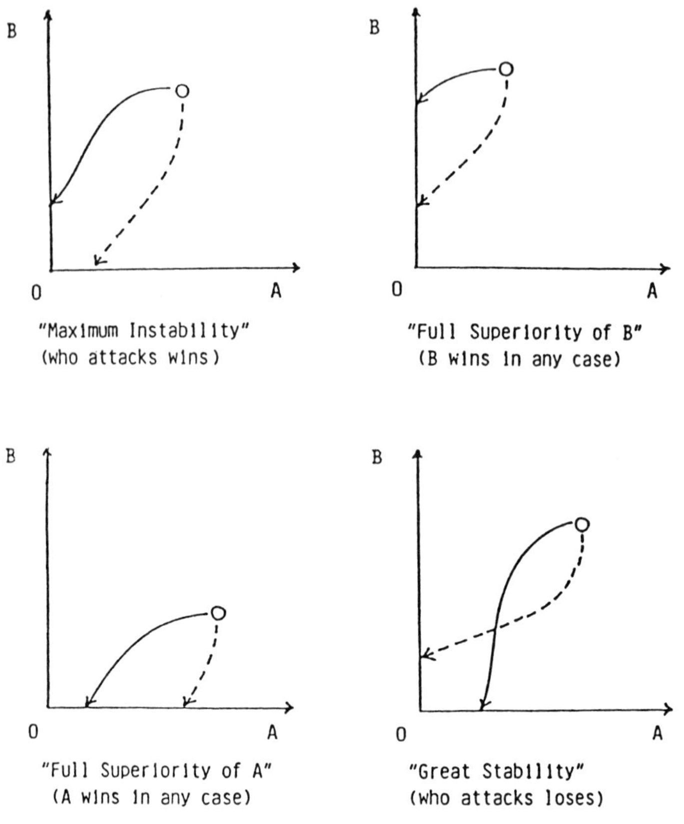

Figure 4 The Four Basic Stability Variants

The arms control dichotomy

In Fig. 5 the top line describes how an arms control agreement may lead to reduced levels of armament, and yet deteriorate the stability features, as the overall configuration is just pushed into the corner, thus losing the crucial intersection point. It follows from this that meaningful arms control in the conventional realm requires an integrated force restructuring process, because only in a situation of a clear mutual defence dominance, can force levels be reduced without harming the stability features of the overall setting.

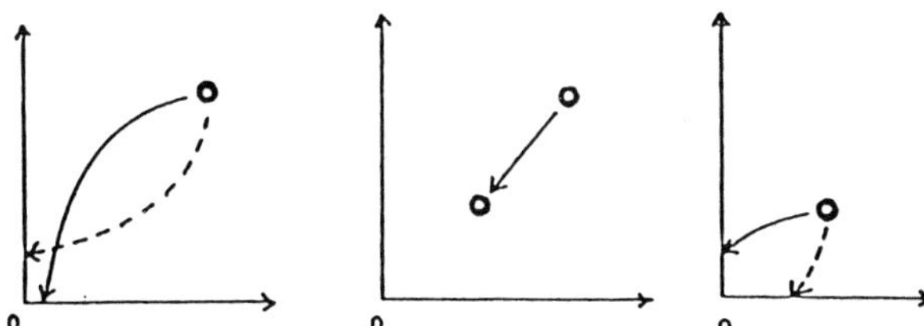

Negative Case 1: Destabilization despite balanced reductions

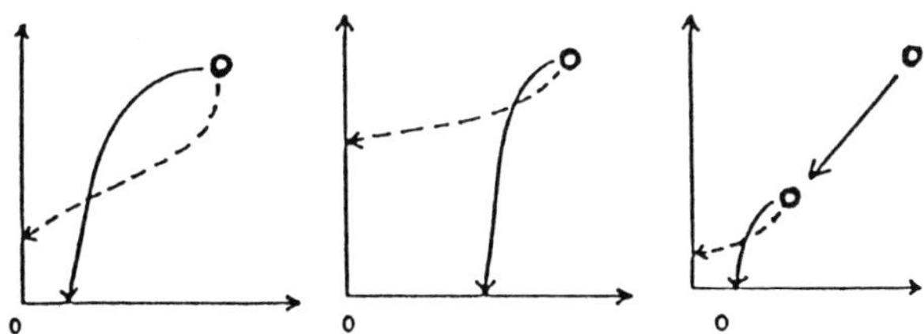

Positive Case 2: Stability at a lowered level of armament

Figure 5 The Arms Control Dichotomy

The decisive conflict of interests: increased military efficiency versus crisis stability (Fig. 6)

Trying to increase military efficiency, military planners often tend to come up with solutions that are very time critical (see, for example, counter-air missions). Those solutions, which may be militarily attractive, at the same time decrease stability as they blow up the "pre-emption belly" (upper chart). Optimizing the stability features of a regime requires exactly the opposite reshaping, namely to reduce the surface of the "pre-emption belly" (which is the surface between the initial force ratio and the intersection point) and to make the final part of the function as parallel as possible to the defender's axis.

The Need for a Much Better Integration of Arms Control Policies and Force Modernization

In order to achieve a regime of a mutual, robust defence dominance, both sides must intentionally reduce their offensive capabilities and tune their equipment, their force structures, and doctrines to the new defensive regime. The decoupling factor between offensive and defensive capabilities

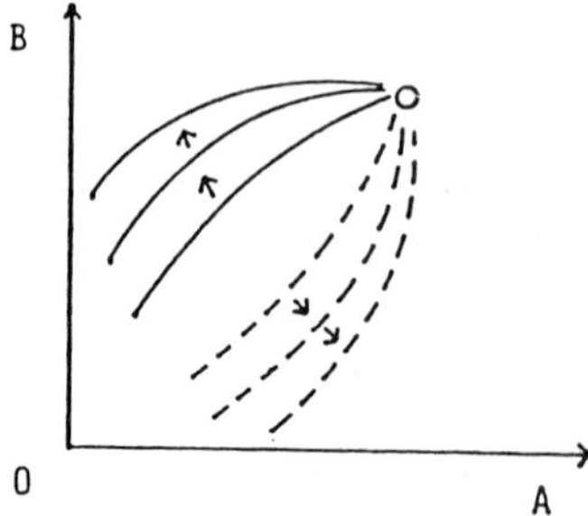

If military efficiency is to be increased inexpensively, measures are often time-critical and have an inherent pre-emptive tendency.

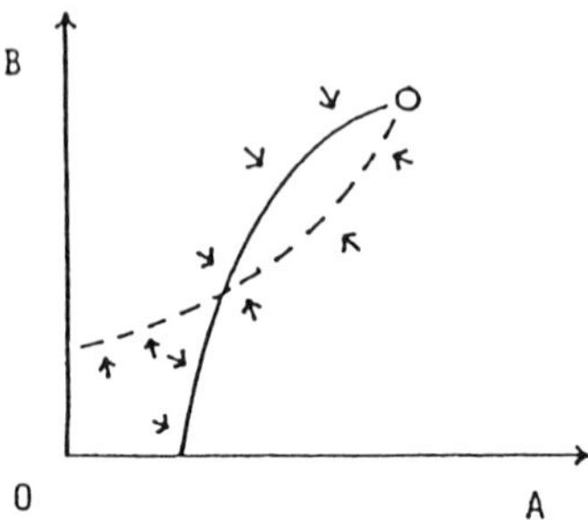

If crisis- and arms control stability are to be increased, preemption bonuses must be reduced and offensive/defensive capabilities decoupled.

Figure 6 The Decisive Conflict of Interests: Increased Military Efficiency Versus Crisis Stability

should be roughly three, that is, the defences of both sides should be capable of coping with offensive forces three times stronger than the opponent actually possesses them. Such a clear decoupling is required in order to compensate remaining pre-emption and surprise effects; to render irrelevant the margins of insecurity when assessing conventional capabilities; and to leave no incentive for a quick build-up of offence-prone equipment and force structures.

A decoupling factor of three can be achieved by a much closer integration of arms control and force modernization policies. Selective reductions of the most offence-capable components (down to low equal ceilings) have to be combined with mutual force modernizations, that is a restructuring process that makes maximum use of the structural advantages of the defender, for example, having mainly local mobility requirements and making terrain preparation.

Outlines of a stability-oriented modernization

In the tables below and Fig. 7 a very rough analysis is given of the technological developments in the conventional realm since The Second World War. And an attempt is made to determine how much progress roughly has been made in these four fields during the last 40 years.

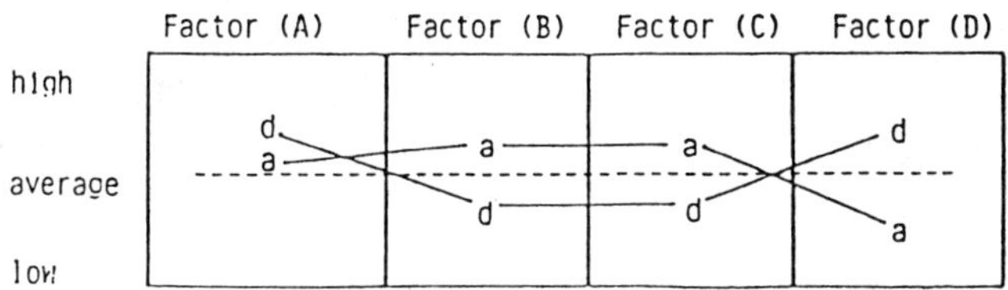

Normal distribution: -----; Attacker: a - a; Defender: d - d;

Figure 7 Resulting Asymmetrical Relevance Profiles

The four basic factors

(A) "Fire" (as explosive and its allocation)
(B) "Mobility" (on land, in the air and in water)
(C) "Protection" (through armour, deception, ECM, etc.)
(D) "Reconnaissance" (early warning and target acquisition)

Technological progress in the above four fields since the second world war

(A) approx. factor 10-100
(B) approx. factor 3-7
(C) approx. factor 3-10
(D) approx. factor 10-1000

The asymmetrical tasks of the attacker and defender

	Attacker's Role	Defender's Role
Advantage	Momentum of surprise and local superiority	Combat on familiar and prepared terrain
Dis-advantage	Forward transport of equipment and personnel	Need for time- and space-covering defence

The identification of the different tasks of attack and defence in the conventional realm (each with one major advantage and one major disadvantage) leads to differing relevance profiles of the four factors. It is argued that the factor "fire" and "reconnaissance" are relevant for both, but are of particular relevance for the defender, since he has to compensate for the momentum of surprise and local superiority by early information and rapid concentration of his own firepower. On the other hand, the factors of "mobility" and "protection" are of special relevance for the attacker as he has to move forward into hostile territory.

Comparing these asymmetric relevance profiles with the progress that has been made in these four fields since the Second World War leads to the following conclusion: there is no automaticity but it should be possible to utilize the over-proportional progress that has been made regarding the factor "firepower allocation" and the factor "reconnaissance" in favour of the defending side. This does not mean that the attacker would not also profit from the progress that has been made regarding firepower and reconnaissance. The argument is that these factors are even more crucial to the basic tasks of the defender and that, therefore, it should be possible to transform the asymmetries regarding technological process into certain advantages for the defending side on a future conventional battlefield.

Outlines of a Stability Oriented Modernization

The most important areas of technology pertaining to "close interdiction"

a) Seismic Sensor Nets in Connection with Stochastically Mobile Mines
- sensor density of approx. 30 per square kilometre
- transmittance via underground fibre glass cables (uninterruptable by artillery shelling and insensitive to EMP and ECM)
- transport of mines by short- and medium-range artillery rocket systems with approx. 3-5 mines per rocket
- search distance per mine is approx. 2000 metres (8 lanes à 250 m)
- approx. 15 kg of explosive per mine
- approx. 15 per cent kill probability per mine
- costs of approx. $3000 per mine
- costs of approx. $50 000 per square kilometre of sensor net (estimate)

b) Simple Reconnaissance and Attack Drones
- data transmittance via fibre glass cable
- transmission range up to approx. 30 km
- remote control by monitor and joy stick
- armour breaking impact on the top and the sides of the attacked vehicles
- multi-sensor technology applied against view obstructions
- costs of approx. $25 000 (large production rate given)

c) Intelligent Passive Munition
- intelligent mines, which scan their environment (up to 100 m) and are autonomously directed and detonated
- stochastically jumping mines, which automatically refill the areas of a mine field that have been cleared already
- communicating mine fields, which detonate only when a certain level of saturation is reached

d) Further Possible Weapon Developments
- shrapnel distribution of small, self-sticking, and magnetically activated transmitters on attacking units, which thus create "cooperative" (easily locatable) targets
- short-term view and sensor blinding through artillery rockets and shells which contain and spray glue on attacking units

Deep interdiction versus close interdiction

Deep Interdiction (as discussed here):	Conventional missiles with intelligent sub-munition against mobile targets in a depth between 100 and 450 km
Close Interdiction:	dug-in seismic sensors; cheap, stochastic ammunition, ranges between 5 and 60 km

	Deep Interdiction	Close Interdiction
A) Military Efficiency	low, as deep inside one's own territory cheap countermeasures are possible (shields, dummies, ECM..)	high, as battlefield constraints apply, extreme difficulties exist in hiding or mimicking seismic signals
B) Cost Efficiency	low, as expensive multi-sensor equipment and steering mechanisms lost with every shot	high, as expensive part of weapon system and explosive are decoupled
C) Crisis Stability	deteriorated, as bonuses for pre-emption are drastically increased on both sides	increased, due to improved forward defence, less targets and incentive for pre-emption
D) Political Signal	highly ambivalent as early or pre-emptive use maximizes effect, no compatibility with arms control; deployment incentive: beef-up first echelon	clearly defensive as no possibilities for an offensive misuse exist; full compatibility with arms control; deployment incentive: withdraw 60 km

The modernization concept of "integrated forward defence"

(1) Build-up of a net of light infantry and barrier units approx. 25 km deep with following provisions:
- force strength in peace approx. 40 000; in a defence situation approx. 120 000 men
- regional draft system and organization in home defence regiments
- reservists' training twice a year on weekends
- mobilization within 12-24 hours possible due to close proximity to defence area and high training level.

(2) A systematic improvement of "close interdiction" capabilities
- laying of approx. 2000 square km of sensor net by members of engineers' corps (costs: approx. $100 million for net, approx. $100 million for supplementary equipment)
- procurement of approx. 3000 simple combat drones (costs: approx. $75 million)
- procurement of further intelligent passive ammunition (guided mines, jumping mines, communicating mine fields, etc. for approx. $125 million)
- procurement of approx. 30 000 mobile mines with launchers (costs: approx. $90 million for mines, approx. $75 million for launchers)
- total costs approx. $565 million, which means price increases even by a factor of 2-3 still tolerable

(3) Maintenance of mobile, heavy-armoured brigades
- with slightly reduced personnel numbers
- not worn down prematurely in the so-called delay battle
- with a high likelihood of being at the right place at the right time
- which can be optimally used according to their characteristics as highly mobile blocking- and counter-attack forces.

The four basic dilemmata

- <u>The Force Level Dilemma</u>
  High force levels: Option for massing and momentum
  Low force levels: Attractiveness for pre-emption

- <u>The Dilemma of the Highly Mobile, Heavy Armoured Units</u>
  Disadvantage: Offence-capable mobility and momentum
  Advantage: Option to compensate offensive concentrations, capabilities for counter-attack and restoration of territorial integrity

- <u>The Dilemma of Ground Attack Airplanes</u>
  Disadvantage: Capabilities for pre-emptive strikes, great relevance of time-critical weapon systems, runway vulnerabilities
  Advantage: Rapid concentration of fire power, defensive tactical air support

- <u>The Dilemma Restricted Military Areas, Zones etc.</u>
  Disadvantage: Possibility of destabilizing mobilization and break-out dynamics
  Advantage: Handicap for surprise attack and offensive massing and momentum
- <u>Conclusion</u>: Conventional Arms Control requires a high degree of systematic analysis and an integrated set of mutually tuned provisions.

Outlines of a Stability Oriented Arms Control Regime

A regime of defence dominance for the land forces (Fig. 8).

Even a perfect equilibrium of today's forces would not provide conventional stability. Ceilings significantly below NATO's present force levels might even decrease stability due to increased bonuses for a surprise attack and blitzkrieg operations. Conventional stability requires the abandonment of the present "all-armour/all purpose" monoculture and, instead, the remaining mobile, heavy armoured units (which are needed for counter attack and restoration of territory) being complemented by attrition-oriented infantry and barrier units (which possess only local mobility).

The military virtues of a combination of heavy armoured, highly mobile units and attrition-oriented infantry and barrier units (Fig. 9).

The attack-capable components of both sides are sufficient for counter attack and restoration of territory. But when crossing the border they are confronted with the same number of heavy armoured, mobile units plus two times as many infantry and barrier units. This regime provides incentives for both sides not to take over the role of the attacker but to remain on one's own territory, because only there can the synergisms with the infantry and barrier units come into play. This holds true even for the most acute political crises.

The attrition-oriented units support the mobile, counter-attack units in four ways: the attacker suffers attrition and thus the force ratio regarding the mobile components shifts in favour of the defender; they gain time thus giving the mobile units a chance to get into optimal positions; they channel the attack and allow the identification of the main axes; they

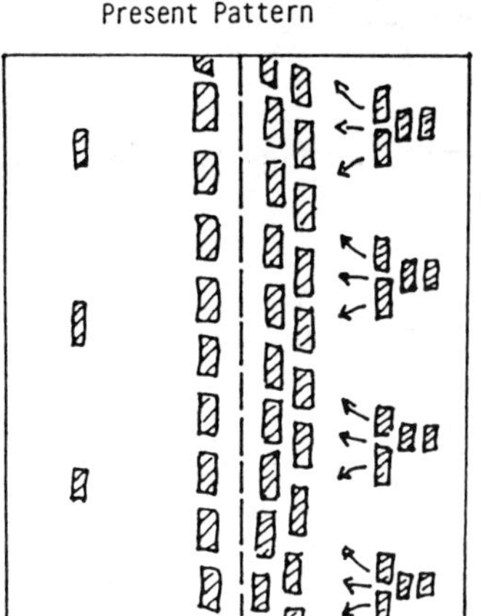

"Invitation for Break-Through"

Force Ratio: 1 to 2.5

▨ : trad. heavy armoured, highly mobile units

⬚ : new, attrition-oriented infantry and barrier units

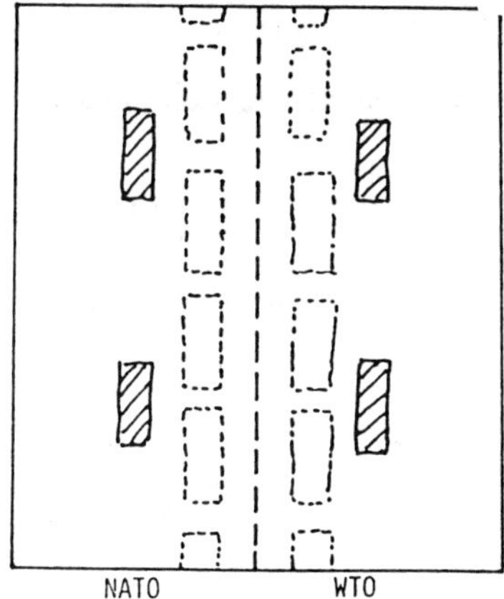

"Defense Dominance"

Force Ratio: 1 to 1

(Equal ceilings for major weapon systems at 50% of present NATO force levels; ratio of heavy, mobile units to infantry and barrier units 1 to 2.)

Figure 8 A Regime of Defence Dominance for the Land Forces

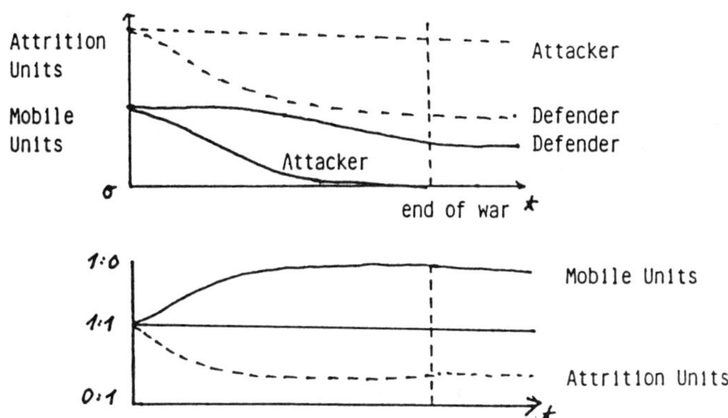

The Development of the Force Ratios

Figure 9 Mobile and Attrition-Oriented Units

create a tactical dilemma for the attacker, because by massing he provides valuable targets for close interdiction and by dispersing he gets in contact with more light infantry and barrier units.

How to avoid the modernization process becoming hostage to the arms control negotiations

There exists a difficult dual task for those engaged on the conventional modernization process: it must make sense, even if arms control agreements will not be achieved. But, on the other hand, it should not block or complicate the latter, but facilitate them.

Modernization Option A:	Modernize and beef up the existing force structures, add deep interdiction capabilities.
Modernization Option B:	Focus investments on new, close interdiction capabilities; complement the heavy, mobile units with attrition-oriented infantry and barrier units.
Effects of A:	An increase of military capabilities, a decrease of crisis stability due to increased time pressures and pre-emption bonuses, no compatibility with a future arms control regime of defence dominance.
Effects of B:	An immediate increase of forward defence capabilities and crisis stability, full compatibility with a future arms control regime of defence dominance (the option to reduce major weapon systems significantly below present NATO force levels is created only by this new combination with attrition-oriented units.

A Stability-Oriented Arms Control Regime in Europe

Substantial Provisions

(1) a low ceiling for MBTs, for example, 10 000) combined with a density limit (for example 500 MBTs per 100 x 100 km)
(2) a comparably low ceiling for heavy artillery (calibres above 100 mm) and rocket launchers, again combined with a density limit that denies offence-capable concentrations;
(3) a low ceiling for strike aircraft (for example, 500) and for armoured helicopters (for example, 500);
(4) a range limitation for conventional rocket and all other un-manned weapon systems (for example, 50 km);
(5) geographical limits on ammunition stockpiles (for example, not closer than 150 km to the border) and no forward deployed mobile bridging equipment at all;
(6) logistical infrastructures that require frequent back-up by non-mobile service stations and other installations.

The Role of Nuclear Forces

For the time being it is probably unavoidable to retain nuclear weapons as a factor of ultimate incalculability. But given a regime of conventional stability, the nuclear component could be substantially reduced.

A limit of 500 warheads for land and air-based systems in Europe might be sensible, of which only 100 would be allowed to be mounted on rockets (with less than 500 km range according to the INF treaty). The arms control proposals for Europe should be complemented by such a nuclear component.

(All limitations apply to the area from the Atlantic to the Urals; regarding (1) and (2) above there is a sub-ceiling of 50 per cent for the Central Region).

A Formula for the Transition Process (Fig. 10).

The transition formula must fulfil two fairly contradictory criteria: it must provide a continuous improvement of the force ratio in favour of the weaker side, and it must provide incentives also for the superior side to engage in arms control. Only substantial progress towards a robust, mutual defence dominance is able to fulfil these two criteria at the same time.

The "Retrograde Ceiling Definition" (RCD) combines this with strong incentives to tell the truth and equal "substantiality" of the required reductions for both sides. The RCD approach works as follows:

- The final, equal ceiling for all weapon categories to be controlled is defined.
- Both sides declare how many elements of each of these weapon categories they possess at present. This defines their initial surpluses.
- By cutting the respective surpluses iteratively three times by half, intermediary ceilings are defined which have to be observed subsequently during the transition process.

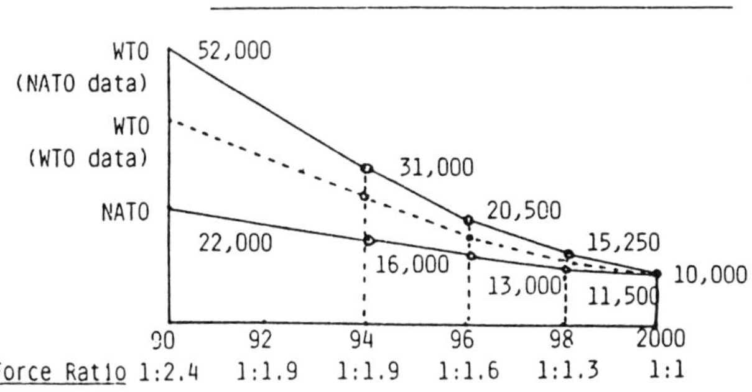

Figure 10 A Formula for the Transition Process

An Overview

The arms control proposal given above is not the only one forwarded up to now. In order to give an overview, we list in Fig. 11 the seven main approaches to conventional arms control (vertical) and evaluate them against the background of six criteria (horizontal). The principals are: (1) to cut units and/or troops and/or weapon systems to the same number (a proposal forwarded by the WTO); (2) to cut these by equal percentages (also forwarded by the WTO); (3) to establish equal ceilings at 95 per cent of NATO's present force levels regarding main battle tanks, artillery and infantry fighting vehicles (a proposal which will probably be NATO's opening position at the CT); (4) the establishment of equal ceilings for active duty forces at roughly 90 per cent of NATO's present force levels, but including aircraft and specific provisions for restricted military areas (a proposal

Criteria / Proposals	Stability I Pre-emption Incentives	Stability II Déf.Dominance (Battle-Field)	Stability III Force Generation potentials (arms race st.)	Savings (NATO/WTO)	Verification Requirements	Acceptance + Implementation (NATO/WTO)
Equal Cuts WTO	- -	- -	0	+ / +	medium	- - / + +
Equal % Cuts WTO	-	-	0	+ / +	medium	- / +
Equal Ceilings at 95% of NATOs MBTs, ARTY, IFV NATO	0	+	+	0 / +	medium	+ + / - -
E.C. at 90% +air craft +RMA/demob. J.D.	+	+	+	+ / +	high	+ / 0
E.C. at 50% (J.D.+close interdiction) A.v.M./G.S.	+ +	+ +	+	+ + / + +	very high	- / -
Air C.ag.Tanks POL	(+)	(+)	(+)	(+ / +)	very high	(- / +)
D.ex/E.C./Def. SU	(0 → +)	(0 → +)	(+ → + +)	(0 → + / + → + +)	very high	(+ / +)

Figure 11 Stability-Oriented Arms Control Regimes

forwarded by Ambassador Jonathan Dean, former head of the US delegation to the MBFR talks); (5) the establishment of equal ceilings at roughly 50 per cent of present NATO force levels for these most offence-capable components, plus limitations which allow only close but not deep interdiction capabilities (this is the proposal described above in greater detail); (6) to reduce Western air attack capabilities in exchange for drastic reductions of Eastern tank capabilities (a proposal brought into the discussion by the Polish government but not yet described in greater detail); and finally (7) the recent Soviet proposal which foresees three stages; (a) the full exchange of data plus verification and the reduction of all offence-relevant weapon categories to the level of the inferior side, (b) the reduction by 500 000 troops on each side, (c) the transition to a posture of clear, structural defensiveness on both sides (this proposal has been forwarded by the Soviet leadership and has been recently backed up by the WTO summit meeting).

Summary

The First Détente was successful regarding the political and economic ties between East and West. But it did not prevent the increase of military threat potentials. Therefore, it was fragile and reversible. It must be the essence of a Second Détente to include a systematic build-down of the military threats in Europe.

A regime of conventional stability is possible but it cannot be achieved as long as the emphasis is on highly offence-capable land and air forces. Conventional stability at significantly reduced force levels requires profound changes in equipment, doctrine and force structures. This, in turn, requires a much better integration of arms control and force modernization policies.

Nobody argues that NATO should make concessions regarding its security. But it would be a mistake of historical dimensions if the West would allow the Gorbachev years to pass by unused, only because of defence parochialism and bureacratic inertia. NATO could and should respond constructively and try to utilize the present thaw for a fundamental and lasting change in the post-war military realities in Europe.

Confidence- and Security-Building Measures in Europe

Adam Rotfeld

Objectives

As defined by the CSCE Final Act, in essence these measures should lead to the following objectives:

- eliminate the causes of tension that exist among states and thus contribute to strengthening peace and security in the world;
- strengthen confidence among states and thus contribute to increasing stability and security in Europe;
- refrain in mutual relations, as well as in international relations in general, from the threat or use of force against the territorial integrity or political independence of any state, or in any other manner inconsistent with the purposes of the United Nations and with the Declaration on Principles of the CSCE;
- reduce the danger of armed conflict and of misunderstanding or miscalculation of military activities which could give rise to apprehension, particularly in a situation where the states lack clear and timely information about the nature of such activities;
- support efforts aimed at lessening tension and promoting disarmament.

In other words, confidence-building measures are to impart credibility to the absence of inimical designs, and bear out declared peaceful intentions rather than introduce curbs on armaments and armed forces. They are more intended to prompt states to show restraint in their military activities than to reduce their military potentials. The continuing tension and crises in international relations have become the main impulse to the search for instruments which would enable states to control and steer the development of the situation. This is especially applicable to East-West relations.

The Measures and Their Implementation

The Helsinki Final Act provided for the following measures: notification of major military manoeuvers (if the number of troops involved exceeds 25 000); possibility of notification of minor military manoeuvers; exchange of military observers (invitations are of voluntary nature and are extended under "mutually acceptable conditions, in the spirit of reciprocity and good will..."); possibility of prior notification of major military movements (this measure has not been used even once since the signing of the Final Act in Helsinki in 1975); other confidence-building measures (invitation of military personnel, exchanges of visits by military delegations, visits by warships, etc.).

Applying these measures in practice during 1975-1985, did not give rise to reservations. The representatives of the CSCE participating states in Belgrade (1977-78) and Madrid (1980-83) did not find any instance of violation of the adopted provisions about notification of major military manoeuvres. It is worth recalling that the authors of the Final Act stated that "this measure deriving from political decision rests upon a voluntary basis". Some NATO states were critical about, what they called, the excessively restrictive realization of these provisions by the socialist states. There were also some reservations - on both sides - about the practical application of the recommendations concerning the invitation of observers. But these were not substantial charges. One thing was beyond any doubt: the provision adopted in the Final Act, which refers to the sphere of military activity, were of a too modest and marginal nature to affect essentially the state of security in East-West relations.

CSBMs and Conventional Reductions in Europe: the Stockholm Document

Of crucial significance to the development of confidence- and security-building measures was the message addressed by the Warsaw Treaty states to the NATO countries and other European nations with a programme for conventional force and armaments reductions in Europe. This document, adopted at the meeting of the Warsaw Treaty Political Consultative Committee in Budapest (10-11 June 1986), embodied a comprehensive approach to disarmament in Europe. The Warsaw Treaty states proposed a significant reduction in all components of ground and tactical air strike forces of the European countries, and the corresponding US and Canadian personnel and material deployed in Europe. Along with conventional weapons, reductions would also embrace operational and tactical nuclear weapons of a range up to 1000 km. The area covered would be the whole of Europe: from the Atlantic to the Urals. The reductions would be carried out in stages, preserving a lowered military equilibrium and involving no encroachment on the security interests of the states concerned. The Warsaw Treaty states thought that conventional disarmament should begin with a single mutual reduction in the sizes of NATO and Warsaw Treaty forces by 100-150 000 troops on either side. Over a period of 5-6 years this would grow to over 500 000 each. In addition, the proposed reductions procedure was designed to make this process lead to a steady lessening of the danger of surprise attack, and to further military and strategic stability on the European continent. It envisaged, for example, a military build-down along the border between the Warsaw Treaty and NATO countries. The Warsaw Treaty proposal provided for the reduction of conventional forces and armaments to take place under effective monitoring conditions, with the use of both national and international systems including on-site inspection. The plan accommodated the frequently formulated demands of the West that reduction and elimination of nuclear weapons in Europe be accompanied by substantial decreases in conventional forces. The proper forum for discussion of the entire range of disarmament issues in Europe would be the second stage of the Conference on Confidence- and Security-building Measures in Europe.

The Document finally adopted by the Stockholm Conference (19 September 1986) comprises six chapters (104 provisions) and four annexes forming an integral part of the agreement reached.

a. Non-use of force

The chapter headed "Refraining from the Threat of Use of Force" reaffirms the commitments in this regard stemming from the UN Charter and the CSCE Declaration of Principles. "No consideration", it states, "may be invoked to serve to warrant resort to the threat or use of force in contravention of this principle" (paragraph 10). The participating states "will refrain from any manifestation of force for the purpose of inducing any other state to renounce the full exercise of its sovereign rights (paragraph 12). The principle of peaceful settlement of disputes is an essential complement to the duty of states to refrain from the threat or use of force, both being essential factors for the maintenance and consolidation of peace and security" (paragraph 17). In this context the CSCE participating states emphasized their commitment to all ten principles. In the Stockholm Document they reaffirmed non-use of force **erga omnis** - "in their relations with any state, regardless of that state's political, social, economic or cultural system and irrespective of whether or not they maintain with that state relations of alliance" (paragraph 15).

Of the ten principles, particular stress was placed on those of sovereign equality, fulfilment in good faith of obligations under international law, and respect for human rights and fundamental freedoms. Effective exercise of these rights and freedoms is an essential factor of international peace, justice and security. This chapter also indicates the necessity to take resolute and effective measures to prevent and to combat terrorism in international relations.

b. Notification

In the chapter dealing with notification it is agreed that notice of military activity, in the zone of application for confidence- and security-building measures, will be transmitted through diplomatic channels 42 days or more in advance.

Military activity by land forces will be subject to notification if it involves at least 13 000 troops or at least 200 battle tanks, or is "organized into a divisional structure of at least two brigades/regiments, not necessarily subordinate to the same division". The term "land forces" is here understood to include amphibious, airmobile and airborne forces. The participation of air forces is to be notified if 200 or more sorties by aircraft (exclusive of helicopters) are likely to be flown. Notification should also include amphibious or airborne operations if they involve 3 000 or more troops.

Also notifiable are transfers of troops into or within the CSBM zone of application if, in size, they correspond to the parameters laid down for land-force manoeuvers (13 000 troops or 300 battle tanks, and so on).

Detailed specification is made of the extent and nature of the information that should be included in notifications.

c. Observation

By comparision with the CSCE Final Act there has been a basic change in the parameters relating to observation of military activities. Observers are to be invited if the number of troops engaged is 17 000 or more. Invitations need not be extended to military activities carried out without advance notice to the troops involved, unless these have a duration of more than 72 hours. The principles of extending invitations, the condition to be met by host countries, and the rights and obligations of observers, have been laid down in detail. Observers are to be granted diplomatic immunity.

d. Annual calendars

A new element are annual calendars of military activities which are to be annually transmitted no later than 15 November (an exception was made for 1986 when the deadline was extended to 15 December).

e. **Constraints**

The Stockholm Document has also introduced certain constraining provisions with regard to military activities:

- if they involve 40 000 or more troops they are to be notified two years in advance (and are prohibited, if not included in the annual calendars);
- if they involve more than 75 000 troops they cannot be carried out, unless notification has been given two years in advance.

f. **Verification**

In monitoring compliance with agreed confidence- and security-building measures a significant role can be played by national technical means (paragraph 64). Participating states are entitled to conduct such inspection of any other state within the CSBM zone. No state is obliged to agree to more than three inspections per year on its territory (and more than one by the same participating state). The Stockholm Document provides for inspection on the ground, from the air or both. In view of the sensitivity of this area, and the lack of experience, the conditions of inspection and the rights and obligations of inspectors have been itemized in considerable detail.

Assessment of the Stockholm Decisions

As this brief recapitulation of the decision of the Stockholm Conference has shown, they incorporate a number of qualitatively new elements in the field of confidence- and security-building measures.

First, the decisions adopted at Stockholm are politically binding and thus, in contrast to the Final Act, obligatory.

Second, they are militarily significant in that they not only make it possible to form a picture of military activity in Europe (notifications and annual calendars) and observe its conduct, beginning at a relatively low quantitative ceiling (17 000 troops), but are also conducive to a limitation of its scale. This is of importance with regard to both major manoeuvres and, to an even greater extent, troop transfers. Essentially, these are new steps hitherto unknown in the practice of relations between CSCE countries, still less between states belonging to opposed alliances.

Third, the zone of application for military CSBMs is much wider than in the Final Act, and it is also worth noting that the extension has been effected entirely eastwards: the formula "from the Atlantic to the Urals" encompasses the whole European territory of the Soviet Union without including even a part of North America, whether the United States or Canada. Furthermore, the measures agreed at Stockholm are not applicable to independent naval activities (unless they form an integral part of land operations in the European continent).

Fourth, there is no precedent in East-West relations for the verification measures agreed. Of particular significance are the decisions relating to on-site inspection which will unquestionably contribute to the greater credibility of the entire system of confidence- and security-building measures. They are evidence of a fairly fundamental change in the Warsaw Treaty states position on this issue. Bearing in mind that this has been a sticking-point in arms limitations, it will be seen that the new control and verification measures should serve as a kind of proving-ground for an agreement on conventional forces and armaments reductions.

The CSCE Vienna Meeting

The agreement in Madrid on the mandate of the Conference on confidence- and security-building measures and disarmament in Europe, as well as the implementation of the decisions of the Stockholm Conference, not only introduced into the agenda of the Vienna CSCE-Follow-up Meeting the issues of continuing the work of that conference, but also created a new platform of negotiations "within the framework of the CSCE process". It was devoted to conventional arms reductions and military stability in Europe. Among the 23 NATO and WTO states there was a preliminary agreement (14 December 1987) that:

> "the objectives of the negotiation shall be to strengthen stability and security in Europe through the establishment of a stable and secure balance of conventional armed forces, which include conventional armaments and equipment, at lower levels; the elimination of disparities prejudicial to stability and security; and the elimination, as a matter of priority, of the capability for launching surprise attack and for initiating large-scale offensive action. Each participant undertakes to contribute to the attainment of those objectives".

Thus, the Vienna decisions have introduced innovative solutions to the CSCE process. In accordance with the Helsinki Final Recommendations (paragraph 65) all the 35 states participate in the CSCE "as sovereign and independent States and in conditions of full equality". All the CSCE meetings should take place "outside military alliances".

Meanwhile, on the basis of preliminary consultations, the delegations of the 23 states (16 NATO, 7 WTO) have started to hold regular contacts in Vienna - outside the framework and independently of the official talks within the CSCE Follow-up Meeting.

The solution agreed upon constituted a compromise between the Polish and NATO proposals. Poland proposed to supplement the Madrid mandate and conduct talks among the 35 states on the reductions of conventional forces and armaments in Europe; this conformed with the earlier provisions, that during the second stage of the Stockholm Conference - beside the confidence- and security-building measures - negotiations should be undertaken on disarmament. The position of France and some of the neutral states were similar and to some extent convergent with this approach. The United States opposed decisively and rejected such a solution. The final compromise in this matter was achieved mainly on the basis of the Swedish proposal.

On 13 May, 1988, the nine neutral countries (Austria, Cyprus, Finland, Liechtenstein, Malta, San Marino, Sweden, Switzerland and Yugoslavia) presented a final draft of the Concluding Document of the CSCE Vienna Meeting. This Document takes into consideration the results of the negotiations achieved, both among the delegations of the 35 states regarding the continuation of the work initiated in Stockholm, and among the delegations of the 23 states on conventional reductions. The link mechanism between the two sets of negotiations required new procedural solutions, different from previous experiences. They were based to a large extent on the suggestions enclosed in the Polish working document.

On the subject of the negotiations among the 35 states the Vienna Document states:

> "The participating States have agreed that negotiations on confidence- and security-building measures will take place in order to build upon

and expand the results already achieved at the Stockholm Conference with the aim of elaborating and adopting a new set of mutually complementary confidence- and security-building measures designed to reduce the risk of military confrontation in Europe..."

The negotiations on conventional reductions and military stability in Europe among the 23 States "will be conducted within the framework of the CSCE process and in accordance with the mandate agreed upon" by the 23 participants.

The Prospects of CSBMs

The following issues are likely to be discussed:

a. the WTO states will reintroduce their proposals to include in the CSBM range the sea areas and air space around Europe, i.e. to bring independent naval and air force operations around Europe within the scope of CSBMs; they will also seek to adopt limits on military activity;

b. the NATO states signalled, in the draft of 10 July 1987, that they would seek to enlarge the Stockholm Conference results by their own postulates, which have been presented but never included in the Conference's final document (this concerns in particular, the regime of verification, expansion of military information, and the scope of on-site inspections);

c. the neutral states will most probably repeat some of their joint proposals, although the present differences of opinion among them are likely to open the way for various expectations. The common elements within the neutral group will be their postulates to limit military activity, lower the ceilings on notification and monitoring of military activity, and recognition of the special role that these states may play in the process of implementing CSBMs.

Quests for agreements on a 'third generation' of confidence- and security-building measures should be accompanied by negotiations on arms reductions in Europe. The concluded INF agreement means the elimination, within the next five years, of all Soviet and American medium-range missiles in Europe. The USSR has proposed the simultaneous scrapping of operational and tactical missiles, and the establishment of a strict system of verification, including on-site inspection. Finally, in the context of the informal talks on future conventional force and armaments reduction negotiations now in progress, the Soviet Union has expressed its readiness to eliminate tactical nuclear devices from Europe.

Plan for the Reduction of Conventional Forces in Europe

The WTO states declared that they are prepared to make greater reductions than NATO in conventional forces in order to remove imbalances and enhance prospects for an arms-reduction agreement. On the other hand, if the West were found to have a military advantage over the Warsaw Treaty States in any particular category of armaments, it would be expected to eliminate this by making cuts (20 000 tanks on the Eastern side for 1400 aircraft on the Western side).

The Soviet official, Lieutenant-General Konstantin Mikhailov, deputy chief of the Foreign Ministry's Disarmament Department, declared on 23 June 1988 that the Soviet Union was willing to eliminate imbalances between the two sides' conventional forces - ground troops as well as non-nuclear weapons and equipment - as the first step in a three-stage plan.

This first step would involve an exchange of "official, responsible data" about the size of each side's forces in Europe, which would enable negotiators to identifiy "imbalances and asymmetries".

In the framework of the second stage - once disparities between the forces had been eliminated - both NATO and the Warsaw Pact should agree to a reduction of 500 000 troops from each side's armed forces.

This would be followed by a third stage in which each side would agree restructure its remaining forces so that they would lose the capability of attacking the other side and would be capable only of defensive operations. The Soviet Union supported also the Polish initiative "to decrease armaments and increase confidence in Central Europe".

The Polish Government envisages the transformation of military potentials into those of a strictly defensive nature, as a set of measures having a political, doctrinal and technical character. The Polish document (15 June 1988) contains a set of new CSBMs which should become the subject of negotiations within the framework of the follow-up meeting to the Stockholm Conference.

Obstacles to Confidence-Building: How Can They Be Overcome?

Klaus Gottstein

Absence of Trust

Past experience seems to indicate that a really significant reduction in the enormous arsenals of weapons now held by the USA, USSR and their allies cannot be achieved unless some confidence is built on both sides that the other side is not plotting to use military force to achieve its political goals, and will not do so in the future.

At present this confidence does not exist, and both sides deem it necessary to maintain considerable forces for deterring aggression by the opponent. It is true that the constant danger of mutual annihilation by human or technical failure, and the considerable economic burden of the present high level of armaments, have created a desire, on both sides, to reduce this level to one of "reasonable sufficiency". However, in the absence of trust in the long-range political goals of the opponent it turns out to be very difficult to determine a level of "reasonable sufficiency" low enough for the removal of the dangers and burdens mentioned above. Hans Rühle, Chief of the Planning Staff in the Ministry of Defence of the Federal Republic of Germany, expressed a widely held opinion when he said in a recent article: "Arms Control Policy can, at most, mitigate the symptoms of East-West antagonism but it cannot remove its causes".

These causes and the resulting absence of trust have, of course, deep historical and ideological roots. The situation is aptly described by Mikhail Milstein, who said at a meeting of the Bergedorf Discussion Circle in Moscow in March 1987: "Each side claims that its military doctrine has an entirely defensive character. ... But we do not believe NATO, and NATO does not believe us. The USA does not trust us, and we do not trust the USA. How long can we afford not to believe each other, considering the developments in the field of armaments? The possibilities of an accidental, undesired use of nuclear weapons are increasing inescapably..."

That mutual disbelief and mistrust exist after several decades of antagonism may not be surprising. If they are to be overcome in the interest of survival of mankind we must study their nature, the arguments used as justification for feelings of fear, hatred and mistrust, and the measures taken by both sides in response to them. We must then investigate which of these arguments and the measures derived from them are based on misperceptions, misunderstandings, misrepresentations and on unjustified worstcase scenarios, and which of them may be considered to be substantiated by well-documented facts. It will be difficult to agree even on this distinction but that is part of the problem, and the reasons for any disagreement on this distinction will have to be included in the investigation.

The final goal for the countries belonging to two different blocs, representing two different social and political systems must be to substitute their present relationship by one of a new type. It has been said for some time that this relationship should be one of peaceful coexistence. Peaceful coexistence, on the other hand, was defined in earlier Soviet publications as a form of continued class struggle in the presence of limited cooperation. This type of coexistence was not sufficient in the past to remove the mistrust which, in turn, was responsible for the continuation of the arms race. A new relationship between the two systems must be found. The only one offering hope for the future seems to be cooperation combined with peaceful competition. Competition will naturally develop not only economically but also socially in those areas where the approach of the two systems to the solution of problems is different. There will also be many opportunities for learning from each other by cooperation. There is enough scope for cooperation, as one realizes when looking at the problems facing humanity in our ages.

On the other hand, history teaches us that goodwill is not enough for solving problems. The obstacles to an honest and efficient cooperation between East and West are a heritage of the past. And they are real. They can only be removed by changing old ways of thinking and old habits, on the basis of careful analysis. This analysis should lead to judgements as to which traditions are good and should be upheld, and which traditions are detrimental and should be changed. A few examples will illustrate what this means.

Causes of Fear and Mistrust

Let us first look at the basis of the arguments for fear and mistrust. Henry Kissinger, in a speech in Hanover, said that Secretary Gorbachev, as a single person, cannot change 400 years of Russian history which was a history of continuous expansion. We must be prepared, he said, for a continuation of this expansionist policy. President Reagan stated, in 1981, that "Bolshevism is a wrong track of history for which there is no justification, a page in the book of history that ought to be torn out". Nevertheless, Reagan is ready to cooperate with the Soviet Union, although he also said recently that his views about communism have not changed.

Statements of this type from the West may be compared with similarly ominous ones from the East. In the resolution of the XXVIIth Party Congress of the CPSU (1986) there is the following proposition: "Imperialism, with militarism as its monstrous prodigy, is representing a growing threat to all of mankind". Vadim Zagladin, First Deputy Head of the International Division of the Central Committee of the CPSU, said in 1987: "Western social order undermines the foundations of the normal existence of humanity. Herein lies the confirmation that this society has almost exhausted the historical justification of its existence".

Thus, on both sides we find the notion of continued struggle even though both sides favour limited cooperation. This notion is dangerous because it lends support to those on both sides who believe that deterrence will remain necessary. The origin of this notion can be traced to the theoretical universalism of communism, which can be seen as a threat to the non-communist world, and to the theoretical universalism of the American Declaration of Independence, which also speaks of the rights of all people, and explains the interventionist component of US foreign policy visible from the beginning of US history. Each side tends to see its own system in its theoretical, idealized form. The shortcomings of daily practice are described as deviations to be corrected, not as the central characteristics of the system. Only the opposing side sees these "deviations" as typical. So Soviet commentators do not describe life in the Soviet Union as a manifestation of typical life in socialism but just as life in "real socialism". Similarly, defenders of Western democracy might describe some negative aspects of life in Western cities not as typical of democracy but as aberrations occurring in "real democracy".

The notion of continued class struggle certainly looks menacing to those who feel that they themselves were singled out as the targets and victims of this struggle. Some ideologists in socialist countries have recently expressed the view that there is no change in the intrinsic aggressiveness of capitalism and that, therefore, socialism cannot give up its enemy image of capitalism. Capitalism, it is said, is inherently incapable of peacefulness, it has to be made peaceful. Statements of this type look disquieting to Western observers, and are not helpful for disarmament efforts. It may be true that such statements seem necessary for those who feel responsible for maintaining the purity of socialist ideology. And in a reciprocal sense, analogous statements about communism may be considered necessary by those who feel responsible for the maintenance of Western democracy and the protection of human rights. They may require an enemy image also. But we should not forget that these enemy images look threatening to the other side. Therefore, it should not be left to imagination what is really meant. The rules of competing with the "enemy" should be clear and unambiguous. It must be our goal to eliminate completely any ideas that this competition might have to be fought out with military weapons. The two systems differ mainly in their social and economic aspects, not so much in their weapons. So it is in the social and economic fields in which they should compete, and in which they ought to cooperate also.

Unfortunately, we are still far from the universal acceptance of this moderation. The principle of mistrust is still valid, as was recently stated when Defence Ministers Carlucci and Jazov met in Berne. General Wolfgang Altenburg, Chairman of the Military Committee of NATO, said in April 1988 that the strictly defensive character of NATO should not exclude the use of nuclear weapons as a political instrument and as "signal" against an aggressor. The new Defence Minister of the Federal Republic of Germany, Rupert Scholz, said that "for a long time the Soviet military doctrine has amounted to making wars possible again by reducing the nuclear potential, thereby making effective again Soviet conventional superiority". In this context, his predecessor, Wörner, pointed out that each year, from 1984 to 1986, the USSR introduced 1600 new battle tanks. On the other hand, Defence Minister Jazov of the USSR, in his speech on Red Square in November 1987, said that "reactionary imperialist circles continue to strive for military superiority. Soviet forces, in alliance with troops of the Warsaw Pact, are always ready to defend socialist achievements and to fulfil their internationalist duty". That is the situation on both sides in the present political world where, as George Kennan put it in a recent article, artificially-created images are considered more significant than realities.

Need to Pay Attention to Perceptions

If this situation is to improve, we will have to pay attention, on both sides, to these images, or perceptions of long-range goals, study their origins, clarify misperceptions, remove misunderstandings, and uncover false images created intentionally for reasons of ideology or psychological warfare. This will require systematic discussions between representatives of the two alliances. It may turn out that some threat perceptions can only be removed by changing the goals themselves. If so, then, in the interest of survival, this will have to be done even if it means abandoning, or redefining, long-cherished ideologies. But if we just feel annoyed, not threatened, by the other ideology, then we should refrain from missionary zeal. For the sake of peace we should be tolerant. Gorbachev and his followers now proclaim a belief in the superiority of communism while yielding a right of existence to capitalism, leaving it to history what the final outcome will be. Western leaders could express their belief in the superiority of Western liberal pluralism, yielding to the Soviet Union the right to experiment with communism as long as others are not forced to take part in these experiments.

If we concentrate our joint efforts on the solution of global problems, fighting environmental pollution, hunger, disease, shortage of resources, illiteracy, rather than each other, then we shall develop a spirit of cooperation which will give less and less importance to differences in ideology. Our common enemy will be those global problems - environmental pollution, hunger, disease, shortage of resources, illiteracy - and we shall develop a feeling of belonging to one and the same alliance in this battle. Our methods may be different, but the goal of our coordinated efforts should be the same: to take care of life on this planet, and to remove, as far as possible, sources of suffering and unhappiness. In our modern age, this cannot be done without profound knowledge, without using the resources of science and technology. Honest cooperation in tackling the problems of mankind will lead to mutual understanding and to the gradual disappearance of unjustified mistrust. It will make peace more secure.

Prospects of Satellite Verification in Europe

Caesar Voute

Introduction

The issue of satellite verification has been on the Pugwash agenda for years. The subject has been discussed at several of its annual conferences, especially in 1984 (Bjorkliden), and 1987 (Gmunden). Moreover, two special symposia, in 1980 and 1983, were devoted to the theme. Attention was given to the verification technology as such and to "the rules of the game".

The satellite verification issue has also retained the attention of other expert groups and was debated at various workshops and conferences in North America, Europe and the Far East, outside the normal Pugwash environment, although frequently with some Pugwashites contributing. Moreover, there has been no lack of scientific publications devoted to the subject. The bibliography on the subject is already quite extensive[1].

Gradually these contributions are having some impact. In the first place, one has to mention the important Pugwash inputs into the United Nations study on an international satellite monitoring agency published in 1983[2]. But the Pugwash influence can also be traced in some national studies on international or regional satellite verification as an option, more or less offically commissioned by the respective governments, for instance in Canada, the Netherlands and Sweden. Nevertheless, one often wonders whether we have come any nearer to the implementation of an international or regional satellite verification facility, or whether the world will continue to rely on the satellite verification facilities of the two superpowers. This is not so much a matter of technology or of investment and operational costs. The technology is already available in several countries, members of the two main alliances, neutral and unaligned. Like in other fields of high technology costs are decreasing as the technology develops and its capabilities increase. The main problem at this stage is that of political decision making. It is on this aspect that the present discussion should centre.

Technological Developments

The technology of earth observations from space continues to develop rapidly. Remote sensing satellites are becoming more and more powerful. Moreover, the number of countries or groups of countries, operating high- to very-high-resolution remote-sensing satellites continues to increase[3]. At this moment, US, USSR, French (since February 1986), Indian (since 18 March 1988), Japanese (since spring 1987), remote-sensing satellites are already orbiting the Earth. China has launched during the last few years several short duration earth observation missions.

The European Space Agency (ESA) will launch its first remote-sensing satellite (ERS-1) in about two years time, to be followed by Canada

(Radarsat). The USA and Japan will start operating a new generation of remote-sensing satellites with improved capabilities in the early 1990's (Landsat-6 and JERS-1). India continues to develop new missions. Very recently, Brazil and China have announced their decision to develop jointly another satellite remote-sensing system, while Indonesia and the Netherlands are still considering a Tropical Earth Resources Satellite in an equatorial orbit.

The French were the first to operate a civilian remote-sensing satellite (SPOT) with a 10 metre ground resolution capacity. In its two years of operation it has already amply demonstrated the military value of the images acquired, as illustrated by many examples published in the media and in specialized journals. The KFA-1000 satellite photographs being commercially marketed by Sojuzkarta (USSR), with a ground resolution reaching 5-6 metres, constitute a next step. As a consequence, the US administration has relinquished in February 1988 the existing ground resolution limitations for the civilian American remote-sensing satellites. The policy directives, published on 11 February 1988, state explicitly that the US government will encourage the development of commercial systems which image the Earth from space, competitive with or superior to foreign-operated civil or commercial systems[4].

Plans are under consideration to have the American Landsat 6, to be launched in 1991, equipped with a 5 m STAR sensor in addtion to the Enhanced Thematic Mapper. Given the positive experience gained thus far with Landsat TM, and especially with SPOT high resolution images, the new US STAR sensor would be a valuable tool for data acquisition with an important news value. Hence the interest of the media in this new instrument. For policy reasons the US administration would prefer to have the STAR instrument developed, implemented and operated by industry. The media are exploring the possibilities to have this sensor exploited using their resources and under their responsibility, with the government exercising overall supervision to protect national interests.

Perhaps most relevant for our considerations is the Canadian Radarsat project, foreseen for 1992. It will be the first equipped with a microwave (imaging radar) instrument programmable for different ground resolutions and look directions, as well as an option for acquiring stereoscopic image pairs. This sensor with its all-weather capability, will be much better suited for verification purposes than the high resolution sensors operating in the visible part of the spectrum, which can only acquire data at day time and under favourable weather conditions. SPOT and the future Landsat-6 are equipped with such sensors operating in the visible part of the spectrum. And even though ERS-1 and JERS-1, the latter combining microwave sensors with instruments for data acquisiton in the visible part of the spectrum, also have all-weather capabilities, they are still less suited for the special requirements of verification tasks.

It should be noted that the Canadian Paxsat-B concept for space-to-ground verification is based almost entirely on Radarsat technology, existing already or presently under development. A number of improvements would have to be added to respond better to specific needs of the verification of an internal or regional agreement on the reduction and control of conventional armed forces in Europe. In the section on political developments more will be said about this project.

The number of military reconnaissance satellites and their capabilities is also increasing. At this moment only the USA and USSR are operating such satellites. The ground resolution of these various systems is not officially disclosed. However, there exists ample evidence that it is in the decimetre order, far better than the 5 metres of the best civilian system available now or in a few years time. From time to time information is made available to the media, based on an analysis of such satellite observations. Recent examples are the explosion in a Soviet factory in Pavlograd, reported on 18 May 1988; the initial confusion about the meaning of new Soviet construction works in Afghanistan, which in the end proved to be galleries for the public and the international press witnessing the pulling out of the first Soviet troops from the country. In both cases US spy satellites were cited by the official US spokesmen as the source for the information.

The USA and USSR both operate reconnaissance ("spy") satellites equipped with cameras, the exposed films being ejected and captured for processing on the ground, as well as automatic sensor systems transmitting the acquired imagery by radio links to ground receiving stations. In addition, these daylight-only systems are supplemented by all-weather microwave sensing systems, in the Soviet case using nuclear power generators included in the satellite payload.

France, with its HELIOS reconnaissance satellite project, will in a few years time join the two superpowers in this field. The French HELIOS system, operating in the visible part of the spectrum and based on current SPOT CCD array sensor ("pushbroom") technology, would have a resolution better than 1 metre, as compared to the 10 m resolution of the civilian SPOT. The next HELIOS-type satellite will probably also carry a high-resolution microwave sensor being developed in the Federal Republic of Germany to provide an all-weather capability.

The rapid development of sensor technology is being paralleled by an equally rapid development of computer-assisted image processing and data handling techniques. A wide range of digital image processing and interactive or automated image enhancement, classification and information extraction systems, is now available. Moreover, much work is devoted to the development of Knowledge Systems ("Artificial Intelligence"), Geographical Information Systems (GIS), and Land Information Systems (LIS) for complex multi-source information handling operations. In combination with scenarios and models describing various types of military build-up of the party being observed, these are used for analysing the probable nature and intentions of the military activities of the other party.

The power and capabilities of digital hardware and software are increasing rapidly, while the costs of computers and peripherals continue to decrease. Soon a situation will prevail when small stand-alone systems for decentralized image processing and data handling will be available on the market, which will be cheaper than the individual digital image frames that are used as inputs for the analytical work. The speed and accuracy of image processing and data handling are also increasing continuously, an important feature for reconnaissance and verification work, requiring near real-time information.

Civilian and military Earth observations from space have much in common, both with regard to the satellites and their payloads, and in connection with image processing and data handling on the ground. The main differences are a greater emphasis on all-weather capabilities for military (reconnaissance and verification) purposes, a greater need for selective, detailed and accurate observations, and the need for very fast information. Furthermore, the military will have a greater need for centralized complex and expensive image processing and data handling systems, while many civilian applications stand to gain by working with smaller, decentralized and user-friendly facilities.

Political Developments

A first phase in the international political debate on satellite verification started in 1981, when the Secretary-General submitted to the Second Special Assembly of the United Nations on Disarmament, without receiving much response, the study of an Expert Group on an International Satellite Monitoring Agency. In 1984/1985 the Parliamentary Assembly of the Council of Europe and the Assembly of the Western European Union WEU referred to satellite verification in reports and resolutions, and Sweden and Switzerland started to explore the interest of other countries for the availability of verification facilities not operated by the two superpowers, but being supplementary to the "national means of verification" of those superpowers[1,2,5]. In those years the situation was characterized, at best, by a lukewarm interest in some of the smaller NATO members and in some neutral or non-aligned countries, and by complete lack of interest, or even a rather strong opposition, from the side of the two superpowers. The discussion took place almost entirely at an informal level between a number of scientists and a small number of parliamentarians.

Since 1986 the situation has been changing rather drastically. The USSR has been dropping gradually its opposition to various verification procedures other than the agreed "national means of verification" of the two superpowers. Some progress has been made in the various arms control negotiations, culminating in the INF agreement signed in December 1987. In these negotiations verification has always constituted a major issue, and gradually the scope of verification was enlarged, now including even intrusive measures, like on-site inspection upon challenge.

Hesitations about the acceptability of independent satellite verification by countries other than the two superpowers continued to prevail in NATO member countries, as illustrated by the rejection in the Netherlands parliament of a draft resolution on a European satellite verification facility in December 1986. Nevertheless, the concept was gaining wider support than in the past.

A study on security policy aspects of outer space by the Netherlands Advisory Council on Peace and Security, commissioned by the Ministers of Foreign Affairs and Defence, recommended in June 1987 explicitly the establishment of a European satellite verification facility and of a research institute on verification technology[6]. In another policy report, this time jointly by the Institutes of International Affairs in France, FRG, Italy, Netherlands and UK, it is recommended again that Europe should establish autonomy in space to safeguard its own security interests. A European satellite verification and/or reconnaissance facility is considered an essential element for this purpose[7].

Meanwhile, the Canadian Ministry of External Affairs commissioned a double study on satellite verification, covering two concepts: Paxsat-A (Space-to-space verification) and Paxsat-B (Space-to-ground verification), both performed by SPAR Aerospace Ltd., the same consortium which is carrying out the Radarsat feasibility and development studies. In June 1987 a Canadian mission visited a number of European capitals to brief the various governments of NATO members and neutral countries about the nature and scope of the Canadian proposals. The Paxsat-A study covers the subject of arms control in outer space in general, while the Paxsat-B study addresses the specific aspects of verifying controls on conventional weapons in one particular area, namely (Central) Europe.

The Canadian concept attempts to overcome some of the institutional, financial, political and methodological issues militating against broader satellite schemes, such as the International Satellite Monitoring Agency (ISMA) proposed by the 1981 UN study. Unlike ISMA, Paxsat was conceived from the beginning as only to be developed within a specific treaty; to be based on a regional context; and in no way have application outside the treaty. Under Paxsat, participants would be parties to the treaty itself[8].

Follow-up studies on the Paxsat concept are currently being carried out to respond to questions posed during the briefing mission of June 1987 and to elucidate a number of details insufficiently covered by the initial study.

Special mention should be made of the so-called "Stokes" report on "Threat Assessment", submitted to the Assembly of the Western European Union. The report stresses the uncertainties about the presumed superiority of the conventional forces of the Warsaw Pact, and recommends that the European NATO partners should develop an autonomous verification capability to supplement that of the USA. At the December 1987 Assembly the Conclusions and Recommendations were referred back for revision to the Committee. Whatever its present status, this report illustrates clearly the necessity of adequate previous verification in all negotiations on asymmetric arms reduction on the European theatre[9].

Of considerable political significance is the decision of the WEU Council to ask the WEU Agencies for Security Questions to study possibilities for verification measures with regard to different arms control fora[10]. In the WEU Assembly debate on 1 December 1987, Mr. van den Broek (Minister for Foreign Affairs of the Netherlands, Chairman-in-Office of the WEU Council) confirmed that the reluctance of the Council to promote a European military space policy, including the implementation of satellite signals intelligence and reconnaissance, is due to financial constraints and is not based on a political no[11].

This study of the Agencies for Security Questions concerns the verification of conventional armed forces in Europe. It includes satellite observations, and takes into account the French HELIOS and the Canadian Paxsat concepts. Like other studies commissioned by the governments of the WEU member states, this report has not been published and is not officially available to the scientific community.

No less important has been the 12 January 1988 Stockholm Declaration of the six Heads of State (Argentine, Greece, India, Mexico, Sweden and Tanzania), urging the establishment of an integrated and multilateral verification system under United Nations supervision. The new Stockholm Declaration of the Five-Continent Initiative has induced Dr. Rolf Linkohr to table, in February 1988, in the European Parliament, a draft resolution calling upon the European Ministers Conference and the European Commission to establish an Expert Group to explore how the European Community could be involved in scientific and technological aspects of arms-reduction verification. Furthermore, the draft resolution urged the European Commission to participate informally, in the Stockholm expert discussions on this theme, and to study the possibilities for the development and construction of a European satellite system for verification of arms reduction agreements. Finally, Ministers Conference and Commission are urged to provide strong political support to the Five-Continent Initiative with regard to the verification issue[12]. The significance of this draft resolution, resides in the fact that it recognizes explicitly that satellite verification constitutes a peaceful use of outer space. Therefore, the European Community, which is only mandated to deal with civilian and not with military matters, could include satellite verification and monitoring in its science and technology policy.

The verification issue has also retained much attention in the June 1988 debate in the United Nations General Conference, where the Netherlands Minister of Foreign Affairs was one of the speakers asking for another UN Expert study on the role of the UN in the verification of arms control agreements.

In view of these various developments it is regrettable that very little progress is being made in Western Europe on a national level. Early this year space policy was subject of a major debate in the British Parliament[13]. Although space policy was approached in an integrated and comprehensive manner, little reference was made to military uses of outer space. The satellite verification issue was not mentioned at all. The Lords had to confess ignorance on defence aspects, mainly because the government did not consider it necessary to brief parliament on these issues.

Verification Versus Reconnaissance

When critically reading the various studies and policy documents referred to in the previous section, one notes that "reconnaissance" and "intelligence gathering" are often equated with "confidence-building measures", and that "verification" and "reconnaissance" are considered tasks which could be carried out by one and the same system operated by a single nation or members of a single alliance. However, the concepts of "verification" and of "reconnaissance" need careful consideration, since they relate to utterly different activities. Combining or confusing them leads to a very basic mistake which may have considerable political consequences.

"Verification" concerns checking wether parties to an agreement stick to their mutual obligations. The verification task, therefore, implies that the parties to an agreement accept the elements to be verified and the techniques and methods used for verification. Verification is a bilateral or multilateral function, implicitly accepted or explicitly stated when parties enter into an agreement.

"Reconnaissance" concerns an action whereby one party investigates what another party is doing and whether this constitutes a threat to the party carrying out the reconnaissance activities. Reconnaissance is, therefore, a unilateral action the main objective of which is to protect ones' security and interests without much consideration for the interests or security of the "adversary", who need not be in agreement with the reconnaissance activities being performed. Hence, "intelligence gathering", or the pejorative expression "spying", are synonyms for reconnaissance. There is also a difference between the technical means required for adequate verification and reconnaissance. In the latter case, it is important to know the exact strength of the adversary, including the question whether the other party is developing a new powerful arm. Reconnaissance, or intelligence gathering, devices are, therefore, defined to observe, identify and describe as accurately as possible a single arm of the other party, e.g. a single missile, airplane or armed vehicle. The objective of verification is to check whether both parties to an agreement fulfil their obligations, e.g. in abstaining from a build-up of armed forces which can be perceived as offensive and threatening to the other part. The total strength and volume of armed forces in a certain area is decisive, not the identity of a single arm or prototype of an arm. Verification devices, therefore, need not have the same accuracy as reconnaissance devices. Detection of change with time constitutes an important function of verification tools. Another important function is their signal function, especially to trigger more intrusive methods of verification, like on-site inspection upon challenge.

In view of the above, one important question to be addressed is whether a verification system should be treaty-specific, as proposed by Canada for the Paxsat concept[8], or whether a multipurpose verification system, as proposed in the UN study on an international satellite monitoring agency, could become a reality[1,2]?

The second important question, with considerable financial consequences, is what should be the technical parameters for an adequate satellite verification system. Could its accuracy be much less than that of the reconnaissance system of the two superpowers while still maintaining a sufficient capability to function as a supplement to the "national technical means" of those superpowers?

The third essential question is whether parties involved in negotiating an arms control and arms reduction treaty agree to specific verification measures, and whether they are willing to define, develop, implement and operate such verification measures in their common interest? Some degree of cost sharing or joint development of verification technology could be included as an option in any agreement on this matter, but should not be considered a must.

Another question is whether parties to an agreement would accept verification services provided by a third party, e.g. a neutral country or group of countries, as was suggested earlier by Sweden and Switzerland, and in the United Nations study?

What Should Be The Role Of Pugwash?

The situation in Europe has changed considerably since Pugwash first discussed satellite verification and monitoring for arms control and disarmament agreements. The main problem to be solved now is the political decision-making process concerning acceptance and implementation of satellite verification as one of the tools for strengthening existing and future arms limitation agreements, often in combination with other verification measures.

This requires a better definition of **verification functions** in order to prevent further confusion with reconnaissance functions. Pugwash could assist politicians in all countries concerned, and in the international or regional intergovernmental bodies, in the conceptual clarification of these terms. This should be followed by a better definition of the technological parameters for verification systems in order to make these as cost-effective and efficient as possible. Complementarity with national means remains a prerequisite.

A second element requiring elucidation is whether it would be more effective to develop treaty-specific verification systems, or whether it would be possible to achieve ultimately an international or regional multipurpose system. Here again conceptual clarification is of the utmost importance prior to discussing technological parameters and "rules of the game". The role of third parties in providing monitoring services should be included in this analysis.

The third dimension - one where Pugwash could also be quite effective - is in facilitating the negotiating and political decision-making processes as such, by providing all concerned with relevant information and advice. Informal consultation between experts from the negotiating parties, including influential personalities from neutral and Third World countries, could be very helpful in this respect. Experience from the past is evidence of this potential Pugwash role.

References

1. W.H. Dorn, "Peace-keeping satellites", Peace Research Reviews, Vol. X, nos. 5 and 6, 1987, Duncas, Canada: Peace Research Institute.

2. United Nations, 1981/1983, "The Implications of establishing an international satellite monitoring agency", New York: United Nations, 1983.

3. C. Voute, "The future generation of resources satellites - An assessment of new sensor developments in remote sensing applications for the exploration of natural resources", United Nations Department for Technical Cooperation for Development, 1987.

4. Fact Sheet, Office of the Press Secretary, The White House, Washington DC, USA, 11 February 1988; see also: Space Policy, Vol. 4, no. 2, May 1988, pp. 165-71

5. C. Voute, "An International and a regional satellite monitoring agency - the European political climate", in: Satellites for Arms Control and Crisis Monitoring, edited by B. Jasani and T. Sakata, Oxford University Press: SIPRI, 1987, pp. 129-44.

6. "Dutch report on space and defence", Space Policy, Vol. 4, no. 2, May 1988, pp. 160-61.

7. Europe's Future in Space, a joint policy report by Clingendael, DGAP, IAI, IFI, IFR, RIIA, London: Routeledge and Kegan Paul, 1988.

8. F.R. Sleminson, "Paxsat and progress in arms control. Canadian research focusses on remote sensing applications", Space Policy, Vol. 4, no. 2, May 1988, pp. 97-102; "PAXSAT concept. The Application of Space Based Remote Sensing for Arms Control Verification", External Affairs Canada, Verification Brochure No. 2, 1987; "Verification Research: Canada's Verification Research Program", ibid, Verification Brochure No. 3, 1987; unpublished briefing viewgraphs Department of External Affairs (briefing of European governments, June 1987).

9. "Threat assessment", Report submitted on behalf of the Committee on defence Questions and Armaments by Mr. Stokes, Rapporteur, Assembly of Western European Union, Document 1115, 2 November 1987.

10. Reply of the Council to Recommendation 448, communicated to the Assembly on 12 October 1987 and received at the Office of the Clerk on 16 October 1987. Assembly of Western European Union, ibid, p. 100.

11. Official Report of debates, Assembly of Western European Union, Thirty-third Ordinary Session (Second Part), Seventh Sitting, 1 December 1987, p. 87.

12. Draft resolution, submitted to the European Parliament by R. Linkohr, February 1988.

13. United Kingdom Space Policy, Report of the Sub-Committee on United Kingdom, Space Policy, Select Committee on Science and Technology, House of Lords, Session 1987-88, 2nd Report, London: HMSO.

Disengagement in Europe (Pugwash Symposium, April 1988)

The Pugwash Council

The December 1987 Treaty on intermediate nuclear forces represents a tremendous opportunity to push towards larger cuts in the nuclear arsenals. The INF Treaty not only has generated increased public expectations about, and renewed political commitment to, nuclear arms reductions; but it also has established three crucial precedents that will make further agreements easier; the complete removal and dismantling of entire classes of modern nuclear weapons; asymmetry in the reductions, wherein the side starting with more gives up more; and extensive and intrusive provisions for on-site verification.

Early agreement in the strategic arms reduction talks to achieve deep cuts in land-based inter-continental ballistic missiles and submarine-launched ballistic missiles seems likely and should be welcomed, but the attention of negotiators should also be extended promptly to the still uncontrolled shorter-range nuclear weapons that most immediately threaten Europe. Nuclear artillery shells and nuclear-armed missiles with ranges below the 500-kiloton threshold of the INF Treaty pose a high danger of nuclear-weapons use in any conflict in Europe. Many of these weapons would explode on the territory they were intended to defend, and their use would probably lead to an escalation to a strategic nuclear exchange that would be unimaginably destructive even after the cuts contemplated in START.

Creating a Nuclear-weapon-free zone or corridor on both sides of the NATO/Warsaw Treaty Organization boundary would remove these weapons from the area where the likelihood of initiation of a nuclear war would otherwise be greatest, and would be an appropriate stepping stone towards the longer-term goal of a nuclear-weapon-free Europe from the Atlantic to the Urals. As a first step towards achieving such a zone, modernization and replacement of such forces should be halted. To be fully effective and adequately verifiable, a nuclear-weapon-free-zone should also be free of dual-capable delivery vehicles.

The radical reductions and withdrawals of nuclear weapons that are to be hoped for would be more readily achievable if the conventional forces confrontation in Europe were made less threatening. Perceptions that the conventional confrontation is unbalanced or characterized by significant potential advantages to the side that attacks first have been an important factor in shaping today's nuclear arsenals and doctrines. It is unlikely that the nuclear competition can really be ended until the conventional confrontation is convincingly defused.

The key to the conventional-forces problem is a combination of deep reductions and withdrawals emphasizing the types of weapons most suitable for surprise attack and offensive operations more generally, such as tanks, ground-attack aircraft, and bridging equipment. Restructured conventional forces at lower levels can provide a highly stable "mutual defensive superiority" in which each side's defensive capabilities so far exceed the other side's offensive capabilities that no attack in either direction could possibly succeed.

Increasingly detailed prescriptions for achieving such a transformation of conventional forces, to which the Pugwash Workshop series on this topic over the past four years has made major contributions, are gaining serious attention among military specialists and political leaders on both sides. The INF agreement's precedents in dealing with asymmetries and solving difficult verification problems offer new grounds for optimism that the kinds of agreements needed on the conventional-forces side could indeed be negotiated. It is in the interests of both sides to push hard towards this goal.

Removing all chemical weapons from central Europe is an essential complement to nuclear and conventional arms reductions in the region and would be a helpful step towards the worldwide ban on chemical weapons that is so badly needed. Use of chemical weapons is abhorrent in itself, as well as a potential escalation link between conventional and nuclear war. Several recent initiatives and statements by leaders and political parties on both sides have underlined the possibilities for beginning in central Europe the process of actually ridding the world of these weapons. Implementation of a chemical-weapon-free-zone in this limited region would provide an important signal to states in other parts of the world as well as practical experience very useful for completing the design of a world-wide ban.

The appalling chemical-weapons attacks in the Iran-Iraq war - blatant violations of the 1925 Geneva protocol - demonstrate most forcefully the consequences of the world's failure until now to augment the protocols prohibition of chemical-weapon use with a comprehensive ban on production and possession. We deplore these attacks and all uses of chemical weapons, while hoping that at least they may shock the world community into assigning a higher priority to dealing with the chemical weapon threat.

The goal of a safer, more confident Europe is most effectively pursued not by arms-reduction measures alone but also through increased East-West cooperation in the economic, technological, scientific, and cultural spheres. We endorse such cooperation not just for its benefits in confidence-building and tension reduction, of course, but because the world's existing and looming problems with respect to poverty, underdevelopment, disease, environmental deterioration, and mismanagement of resources, like the danger of war, threaten all nations and can only be mastered if the joint efforts of all nations are harnessed cooperatively to find solutions.

The non-military problems that command joint attention link North and South as well as East and West, and their assessment and solution must be interdisciplinary as well as international. Few existing institutions are both international enough and interdisciplinary enough to address these problems effectively. The International Institute for Applied Systems Analysis near Vienna and the United Nations University in Tokyo are among the exceptions; newer examples are the Global Challenges Network centred near Munich and the International Foundation for the Survival and Development of Humanity with its headquarters in Moscow. We hope these innovative institutions flourish and multiply, and that they find effective ways of delivering their message on global thinking and the realities of interdependence worldwide.

Smaller nations have played important roles in addressing both the military and non-military aspects of international security and related global issues, and they should be even more active in these respects in the future. Their roles include unilateral, bilateral and multilateral initiatives for communication, tension reduction, and confidence-building; the exertion of moderating influences from inside and outside the NATO and WTO alliances; and the encouragement and hosting of international organizations for regional and global problem solving.

Part VI

Military Research and Development

Roles of Technological Innovation in the Arms Race

Theodore Taylor

Introduction

Technological innovation can play two different roles in arms races. The first, and most generally acknowledged, is to accelerate them by providing new types of weapons. The second, now becoming increasingly important, is to assist in stopping arms races and accelerating and maintaining disarmament. Examples of the second kind of innovation are new concepts for verification of arms control and disarmament treaties, and new concepts for peaceful means to increase national and global security that can help shift attention from primary reliance on arms. This paper discusses both of these roles.

There is a fundamental difference between destructive and constructive technology. Destroying something is generally much easier than constructing it - a building, a factory, a work of art, an institution. Once conceived and developed, destructive technology generally spreads more easily than new constructive technology. Nuclear technology capable of destroying civilization in a few minutes is in place, and requires a tiny fraction of human effort and resources that was needed to produce the civilization to destroy it. Permanent safe disposal of the radioactive wastes (a constructive technology), from military plutonium production reactors, is likely to cost much more than it cost to make the plutonium in the first place. (Senator Glenn concluded from recent DOE studies that total cleanup costs are likely to exceed $100 billion. This is much greater than the several tens of billions of dollars spent on producing the military plutonium and tritium in the US warheads stockpiles[1].) Nuclear technology for peaceful purposes is floundering in the countries with the largest numbers of nuclear weapons.

Technologies for weapons and for constructive applications are both open ended, in the sense that room for important innovation is unbounded. The history of development of specific types of technologies may follow an "S-curve" pattern: an initial stage of basic conception, development, and limited use; followed by a period of rapid growth in which basic limits have not yet been approached; and a plateau at which basic limits have been approached. But innovation can often lead to development of a new S-curve for a technology that makes use of some new principles while achieving the previous overall type of objective. It is the process of getting started on new technological S-curves that is the main focus in this paper.

Requirements for Successful Technological Innovation

For a new technology to start its steep climb in a S-curve, the following are required:

- people with innovative talents;
- mandate to the innovators (self-supplied or coming from others);

- resources needed to support the innovators in exploration of new concepts (money, facilities, technical and administrative support people, and access to relevant technical and economic information);
- potential users of the fruits of the concepts who will press for actual development;
- resources needed for development after the initial stage of conception and assessment.

These requirements are often interdependent. Substantial initial resources and a strong mandate are often needed to assemble and support innovative technical people before they begin serious exploration of new concepts. Many new concepts have failed to reach the stage of intensive development because of a lack of a user constituency or needed resources, rather than as a result of technical flaws in the concepts. Limitations in the flow of information, caused by military or industrial secrecy, can severely limit access of innovators and developers to constructive interaction with technical people who are not within the secrecy veil.

Innovation and Nuclear Weaponry

The interplay of the above requirements for successful technological innovation can be illustrated by the history of nuclear explosives.

Two "generations" of nuclear explosives have approached the plateau regions of S-curves in terms of: total yield per kilogramme of explosive; minimum dimensions and total weight for substantial explosions; efficiency of fission or fusion of nuclear fuels; and total number of nuclear warheads in the stockpiles of the USA and USSR. The first generation is fission explosives, including those "boosted" by neutrons released by small amounts of thermonuclear fuel in their interior. The second is thermonuclear explosives that require fission explosions to initiate reaction of relatively large quantities of thermonuclear fuel.

The first S-curve started developing soon after the discovery of fission in 1939, although possibilities for self-sustaining nuclear chain reactions had been already thought of by Szilard and others. The Manhattan Project formally started in 1942, and the first atomic bomb design group was assembled at Los Alamos in the spring of 1943, but some important work on the bomb was done between the time of Einstein's 1939 letter to President Roosevelt and that of early 1943[2].

The star-studded array of scientists and engineers assembled at Los Alamos during the next two years had a central objective: to design and build at least one type of fission bomb that could be delivered by a heavy bomber as soon as possible after the needed highly enriched uranium or plutonium became available. The objective was met slightly more than two years later by building two types of weapons: the gun-type uranium bomb exploded over Hiroshima, and the implosion-type plutonium bombs tested at Alamogordo and exploded over Nagasaki. Athough many more advanced concepts were invented during those years than were actually incorporated into these devices, both designs were chosen to have the best chance that they would "work", rather than represent any attempts to push the technology to basic limits. Both bombs weighed about 5 tonnes, and released the equivalent in total energy of roughly 20 thousand tonnes of high explosive.

Throughout this period, all the above requirements for successful technological innovations were generously met, and included the added sense of urgency from wartime fears that other countries might get the bomb first.

These requirements, without the sense of wartime urgency but with a new concern about staying well ahead of the Soviet Union (especially after the test of its first fission weapon in 1949), were also met in the further successful innovation applied to fission explosives that produced dramatic improvements after the end of the war. These involved pressing basic limits to minimum total weight and dimensions, and maximum efficiency of the contained fissionable materials. By 1952, a second US weapon laboratory, at Livermore, California, was also working under all the conditions required for successful innovation, and developed important new concepts.

By the late 1950's, innovation was beginning to produce much less dramatic improvements in fission explosive perfomance than during the previous decade. The S-curve was approaching the plateau set by basic constraints. Minimum total explosive weights had been reduced to less than 50 kilogrammes; minimum diameters to less than 20 centimetres; maximum yields in large fission bombs to more than 20 times the yield of the Nagasaki bomb; boosted efficiencies approached the maximum allowed by basic design principles and constraints[3].

Thermonuclear ("second generation") explosives had a somewhat similar history. Edward Teller pressed for vigorous work at Los Alamos on the "Super", in parallel with the work on atomic bombs, but was overruled[4]. After the end of the war work on the Super gradually intensified, largely due to Teller's persistence. By the time I started working at Los Alamos in November 1949, I found that most of the attention of the notable physicists and mathematicians at Los Alamos - Teller, Ulam, Everett, Mark, and such consultants as Gamow, Fermi, von Neumann, and many others - was on how to make a Super work, rather than on how to improve fission explosives, a relatively unexciting job. Hans Bethe, who was a very active consultant, did not share this enthusiasm for pursuing the Super. A short time later President Truman called for a crash programme on the H-Bomb, and all conditions necessary for the **attempts** at successful innovation were again generously satisfied (except from Teller's perspective). These attempts continued to fail until early 1951, when Teller and Ulam conceived the way to make the Super work, which Oppenheimer subsequently called "technically sweet". Development from then on was so rapid that the first H-Bomb was tested at Eniwetok Atoll less than two years later. Its yield was nearly a thousand times greater than the first fission explosive.

Important advances in use of this basic concept took thermonuclear weapon technology into the steep part of its S-curve through the remainder of the 1950s. Total yields per unit weight increased to several kiloton per kilogramme in the larger weapons - more than 500 times greater than the first fission bombs. Minimum weights and dimensions of weapons in the several hundred kiloton category dropped sharply, making it possible for existing missiles to carry many warheads in one package. By the early 1960s, however, these types of performance indices began to enter the plateau region of the S-curve, as attention shifted to accommodating nuclear warheads to more and more severe operational military requirements.

This does **not** mean that dramatic innovations in nuclear weaponry are no longer possible because they have approached basic limits in their performance.

Serious attempts started in the early 1950s to develop pure fusion explosives not requiring a fission "trigger". These attempts have apparently not yet succeeded. They are related to so-called "inertial confinement" fusion, under development for many years as a possible form of nuclear power. But pure fusion weapons would require much more compact means for causing small units of thermonuclear fuel to explode with some reasonable efficiency. Conditions for successful innovation have not been as generous as those for the first two generations of nuclear explosives; they are not subject to any announced crash government programme, as were the other two. But major efforts continue.

Perhaps more important are efforts, some of which started at low levels of support and government enthusiasm in the late 1950s, to develop nuclear explosives for which specific forms of energy are greatly enhanced, and others sometimes suppressed. An example of such a weapon that has been developed up to deployment status is the enhanced radiation (neutron) bomb. Many of these concepts also introduce means for directing the selected energy in a particular direction through narrow cones. The number of significantly different possibilities is practically countless. Nuclear explosives release - or can be designed to release - gamma-rays, X-rays, neutrons, radioactive materials, plasmas, extremely high velocity small pellets, visible light, and electromagnetic radiation pulses with wavelengths from the microwave region up to many kilometres. By appropriate coupling with their explosion environment, such forms of energy can be enhanced or suppressed, sometimes in particular directions, and can also be converted to energy in shock waves in air, water, or ground, that can also be significantly directional[5,6].

The most publicized "third generation" weapon is the nuclear X-ray laser for possible use as a space weapon against ballistic missiles or ballistic missile defence systems in space.

Work on some of these new types of nuclear weapons has intensified as the US programme on the Strategic Defense Initiative has gathered momentum. I have no direct knowledge of the extent to which the SDI programme has regenerated the kinds of stimuli for vigorous innovation that existed at the US weapon laboratories in the course of development of the first two generations of nuclear explosives. However, I suspect that the environments are not as conducive to wide-ranging innovation as they were when I was at Los Alamos. I certainly hope not.

The main reason I strongly support an immediate stopping of all nuclear tests by the United States and Soviet Union, and an agreement on a Comprehensive Test Ban Treaty as soon as possible thereafter, is because this would ensure that none of these new nuclear weapon concepts will be the start of the steep part of a new S-curve. I have been convinced for years that new types of nuclear weapons can only make the world more dangerous for all of us. But, insane though they are, the races for new types of nuclear weapons continue.

The Roles of the Weaponeers in the Arms Race

I shall use the word "weaponeer" to mean the innovators who invent, or play major roles in inventing new weapon concepts. My remarks are largely based on the experience I had while working on the design of fission weapons at Los Alamos in the 1950s.

My thesis is that the principal drivers of modern, qualitiative arms races, as opposed to races to acquire larger numbers of new weapons that have been developed, are the weaponeers. It was the weaponeers, albeit with only rudimentary ideas about how to make the materials for atomic bombs and the bombs themselves, that started the chain of government actions leading to the Manhattan Project. It was Edward Teller and a few weaponeer associates that led to the decision to proceed with the development of the hydrogen bomb. He apparently also played a major, and perhaps crucial role, in convincing President Reagan that his dream of an effective defence of the USA against massive nuclear attack could become reality.

On a much smaller scale of importance, I played a similar role in Los Alamos. Most of the innovations I was involved in (along with many others there at the time) were not in response to firm requirements for new types of weapons; they were the result of intense but rather free-wheeling examinations of many different possibilities. As some of the concepts became likely candidates to expand the use of nuclear weapons for both strategic and tactical purposes, I spent more and more time "selling" them, not only to my superiors at the laboratory, but also to people in the US military establishment. Successful "sales" by weaponeers led to formal authority to proceed with nuclear tests and, if warranted, development to enter the national stockpile of weapons. The entire process of pressing harder towards extreme technological limits, sorting out different concepts, and then selling those that looked most interesting, was exhilarating in the extreme.

Over the years since I left Los Alamos I have come to view weaponeering as a symptom of an addictive disease somewhat similar to alcoholism or addiction to drugs. Like the others, the disease is incurable. The only effective control is total abstinence.

I am unaware of any study of the validity of this thesis as it may apply to weaponeers in the fields of chemical and biological warfare or other non-nuclear means for mass destruction. But I strongly suspect that it does.

Technological Innovation Applied to Verification of Arms Control and Disarmament Treaties

With some relief, I now turn briefly to the first of two types of constructive technological innovation - new concepts for verifiying compliance with treaties for the purpose of slowing down the arms races, or actually eliminating weapons. Since this is potentially a huge field, I shall limit myself to a few comments and examples appropriate for verifying agreements concerned with nuclear weapons.

Innovators in this field often find themselves subject to strong intellectual stress. On the one hand, they must give attention to concepts for verifying that nations are, in fact, carrying out their obligations under a treaty, or less formal agreement, i.e. proving that what is believed to be happening is actually happening. The spirit of this is often more confirmatory than adversial. On the other hand, the innovator often is searching for possible ways to cheat, and then devising methods for revealing such cheating. Some of us dealing with these two sides of the issue are often frustrated, especially when we find ourselves better at cheating than revealing it.

Among the many challenges to innovators in this field are the following:

- further decreasing the threshold for detection and identification of underground nuclear explosions, especially via high frequency components of seismic signals and, possibly, via eletromagnetic signals;

- sealing, tagging, and internal "fingerprinting" of identified components of nuclear-weapon systems to be eliminated, including warheads, to assure no substitutions, without revealing information that is nationally classified;

- verifiable, safe, and economical techniques for eliminating components of nuclear-weapon systems, including warheads and special nuclear materials in forms usable in warheads;

- methods for distinguishing between nuclear and conventional warheads for dual-purpose delivery systems;

- inspection of space payloads, without unacceptable interferrence with space launches or operations in space, to assure they contain no weapons or other components, such as nuclear power supplies, that are forbidden by the treaty;

- inspection of large vehicles, such as naval ships, to assure they have no hidden nuclear weapons;

- development of on-site and remote techniques for verifying specified halts in production of components of nuclear-weapon systems, including special nuclear materials;

- development of means to disclose hidden stockpiles of forbidden nuclear weapons or their components, especially nuclear warheads or special nuclear materials.

Innovative approaches to these and other verification and detection techniques are increasingly under study or development by both government and non-government organizations. Examples of the latter types of studies are those under way in the joint project of the Federation of American Scientists, and the Committee of Soviet Scientists for Peace and Against the Nuclear Threat, and the Pugwash-sponsored book on verification. But much more needs to be done in the future than has been done in the past.

Many of the people and resources now used for weaponeering are especially appropriate for new innovative efforts to control and eliminate weapons. They can play key roles in shifting from moving weapon S-curves to their steep climb to rapidly moving S-curves of weapons of mass destruction, to their stages of steep **descent** to the smallest possible plateaus, perhaps to zero. In such efforts former weaponeers from different countries can become colleagues instead of competitors.

Shifting from Weaponeering to Intensive Constructive Action

One of my colleagues, Lewis Bohn, has given much attention to what he calls "Intensive Constructive Action." We, and many others, are pressing for shifts of global attention away from maintaining weapons of mass destruction as principal guarantors of national security, to dealing promptly with the vast array of truly ominous **non-military** threats to **global** security.

We humans are poisoning our habitat - the earth's air, its water, its soil. We are beginning to cause global instabilities in climate by releasing vast quantities of "greenhouse gases" - carbon dioxide from the combustion of fossil fuels, freon, and others - while at the same time destroying forests that help maintain a global carbon dioxide balance.

These and many other large regional and sometimes global disturbances of our habitat threaten to overwhelm our capacity to meet our basic biological, physical, and cultural needs. At the same time that this capacity of our civilized habitat is threatened, the population explosion continues at alarming rates in many regions.

In short, trends in human activity that have reached global proportions in the 20th century now threaten the security of all of us, prosperous or poor, obese or starving, educated or illiterate, living in luxury or homeless, free or oppressed. Weapons of mass destruction pose only one of the threats. It is necessary, but not sufficient, for them to be eliminated if civilization is to survive well into the next century. We must also find ways to come to harmonious terms with our habitat, as well as with each other.

We need a global call to innovative, intensive, and construtive actions to meet these two challenges, if our visions of the future are to be filled with hope and not despair.

References

1. Statement by Senator John Glenn on "Enormous Department of Energy Nuclear Facility Cleanup Cost", 1 July 1988.

2. R. Rhodes, The Making of the Atomic Bomb, Simon and Schuster, 1988, pp. 394-485.

3. T.B. Cochran, W.M. Arkin, and M.M. Hoenig, Nuclear Weapons Databook, Volume I: U.S. Nuclear Forces and Capabilities, Ballinger, 1984.

4. Rhodes, ibid, p. 539.

5. Theodore B. Taylor, "Endless Generations of Nuclear Weapons", Bulletin of the Atomic Scientists, November 1986, pp. 12-15.

6. Theodore B. Taylor, "Third Generation Nuclear Weapons", Scientific American, April 1987.

Controlling Military Research and Development

Jack Ruina

Introduction

In our profound desire to get on with arms control, it is natural to consider limiting military research and development - the conceptual and early gestation stages of military hardware. In our eagerness, we may overlook the complex issues that make control of military R&D generally not feasible and sometimes not desirable even from an arms control perspective. This paper attempts to elucidate why "to check military R&D" as such is so problematic. Although the paper offers no new arms control proposals, it should provoke discussion of what is and what is not realistic in trying to control military R&D.

Addressing the Symptoms Rather than the Cause

A recurrent theme within the communities interested in arms control, and particularly at Pugwash, is that it is essential to impose broad limits on weapons innovation. This is because the arms race in the industrial nations seems as much the pursuit of modernization as the pursuit of greater numbers of weapons. The continuous acquisition of nuclear weapons at a rate beyond any reasonable military or political justification is believed to come not only from deep-seated political hostility and distrust but also from a constant stream of innovations in weapons technology.

Political hostility and distrust undoubtedly lead to the primitive feelings that more and better weapons provide more security, that it is better to err on the side of acquiring too many weapons than too few, that it is necessary to deal with an adversary from strength, that the best way to maintain peace is to prepare for war - and so on.

Looking at the technological factor, military research and development is thought to be a mindless pursuit, with a life of its own promoted by a politically and economically powerful constituency. The products of military R&D present an array of possibilities that military and political leaders can hardly resist; even preoccupation with new military technology arouses potential adversaries' fear and distrust.

Attempts at nuclear arms control to date have emphasized limits on numbers, including the total ban of certain weapons. On the other hand, military research and development has hardly been constrained by international agreement, or by unilateral actions stemming from concern about technology exacerbating the arms race (although the Limited Test Ban Treaty, the Threshold Test Ban Treaty, and the ABM Treaty have restricted testing of some weaponry).

Although there is obvious synergism between the pace of technical innovation in military hardware and the level of political suspicion and fear, the two factors clearly do not have equivalent roles as stimuli to the arms race. We can well imagine a continued preoccupation with arms and arms acquisition even without a high rate of technological innovation - it does not take much to get governments interested in weapons when the political climate is "right" for that purpose. But without political hostility, it is hard to see how an arms race could be sustained by a technological push.

It is not at all clear how much leverage contraction of military R&D would provide in moderating production and deployment of weapons. It is true that R&D increases the menu of possibilities in military hardware, but without major innovations stemming from military R&D, enough possibilities would still exist for sustaining much more of an arms race than we would like.

In a sense, trying to limit R&D is in itself a kind of technological "fix" to a fundamentally political problem;it addresses what is more a symptom than the cause of the arms race. But, as in many instances of medical care, minimizing symptoms may be part of a treatment regimen with the recognition that it is the underlying disease that requires primary attention.

In considering the imposition of constraints on military research and development, we must realize that, even with the recent improvements in US-Soviet relations, effecting **any** constraints whatsoever on R&D requires dealing with some fundamental conceptual, political, and technical problems. And imposing **broad** constraints is probably not feasible and may not even be desirable, since not all technical innovations in weapons are destabilizing and troublesome - some innovations unquestionably lower the nuclear danger.

Civilian vs Military Technology and the Question of Verification

In general, basic technological advances are applicable to commercial as well as to military hardware. In such basic technical areas as electronics, computer hardware and software, materials, communications etc., commercial and military interests substantially overlap. Only as we move up the spectrum of R&D activities, from research to exploratory development to prototype development and to full-scale testing and production, does the separation of military from commercial applicability become clear, and sometimes not even then. For example, the operating commercial DC-10 aircraft has been readily modified to be a military tanker and at one time was seriously considered for use as a cruise missile carrier. MIRV technology stems directly from civilian space efforts and only at the prototype development stage could it have been clearly recognized for what it was. And, of course, much civilian and military space technology borrow from each other quite directly.

There are indeed certain technical devices where, even in the early stages of the development cycle, the military direction is recognizable as such - nuclear pumped X-ray lasers being a clear example - but these are exceptions. In general it is only at the stages of prototype development and full scale testing when military hardware can be identified for what it is and when constraints can be considered at all.

When it comes to the matter of **verification**, we see much the same problem as in the matter of separating civilian from military technology. Early stages of R&D generally do not require special large facilities and are, therefore, not observable by national technical means, nor even by feasible cooperative procedures. There are, of course, some obvious exceptions, such as basic research utilizing large ground-based or space-based telescopes and large particle accelerators. But in most cases, only at the stages of full-scale testing, production, and deployment is there much chance of finding acceptable verification procedures. Even then, verification of limits on full-scale testing may only be possible for physically large, readily observable vehicles like ICBMs, aircraft, ships, or devices that produce distantly observable unambiguous physical effects, the obvious example being nuclear explosives. Here again there are exceptions, particularly for space-related technologies where some exploratory work can be observed and perhaps broadly identified, but even then it is quite likely that there would be a great deal of ambiguity in exactly identifying what is observed.

Contrasts Between Conventional and Nuclear Weapons

Conventional weapons are designed and acquired for actual battlefield use; marginal differences in performance can decide who wins or loses a battle. Quality can compensate for quantity and can be a very cost-effective way of reaching desired levels of capability. If indeed we concede a legitimacy or inevitability to the maintenance of armed forces - and what other choice do we have today? - can arbitrary limits be imposed on the quality of weapons provided to these forces? If technology can make military radios more reliable, tanks less destructible, aircraft more manoeuverable, missiles with greater ranges and greater accuracy, is it conceivable that any nation would deny the very military it created such possibilities if they make economic and military sense?

This is not to say that certain areas of military technology should not be limited as much as possible - biological and chemical weapons are an example - and that constraining even the early stages of this R&D would not be important. But even in these areas we should note that technological innovations are made possible at least as much by non-military R&D as by military R&D, so that here too truly effective limits can be imposed only at the production and deployment stages.

In the case of **nuclear** weapons the importance of modernization is quite different. The capability of weapons to deter a nuclear attack is not sensitive to marginal changes in technical performance. It is only as nuclear weapons are seen to have roles beyond that of simple deterrents that performance details become increasingly important. If nuclear weapons are to be used to deter all conceivable non-nuclear provocations, to be used at many presumably controlled levels of violence, and to fight and win protracted wars, then interest focuses on tailoring weapons for the utmost performance for each imagined scenario of nuclear conflict. It is the responsibility of the political leadership to reject fanciful and contrived scenarios of nuclear confrontation. Fortunately, political leaders worldwide more and more accept that nuclear exchanges once started are not likely to be controllable, that nuclear wars cannot be won, and that the single justifiable role of nuclear weapons is to deter nuclear attack. If (hopefully) this trend continues, the market for the adoption of new nuclear weapons technology should shrink naturally. Nevertheless, imposing constraints on the deployment of destabilizing systems could reduce our present preoccupation with nuclear weapons.

Research and Development as a Safeguard

The history of arms control since World War II shows great hesitancy to agree on any meaningful arms limits. Fear persists that unobserved treaty violations or quick treaty withdrawal would allow an opponent to deploy rapidly hitherto limited systems, thereby giving him a substantial military advantage. An R&D programmme in those very technologies that are limited is considered a safeguard against such eventualities. Continued R&D allows for retention of a cadre of experts in the relevant technologies and for rapid movement to deployment if called for. It gives greater confidence in understanding the potential of new technological developments and their applicability to the weapons areas that are limited. Also, the more data and expertise R&D provides the better the ability to judge an opponent's activities. For these reasons, most US defence science specialists who oppose the US SDI programme and any ABM deployment nevertheless support continuation of ABM R&D.

There is a general sense in the USA that technology is our "strong suit" and must be vigorously pursued both for commercial and for national security purposes; harm ensues only if technology is badly applied. Although the military services are quite prepared to consider production and deployment limits, they see R&D as providing some of the required safeguards against unanticipated technological innovations. Thus, in general, attempts to limit R&D would face strong domestic resistance.

Apropos of the question of the role of technology in the arms race, we can cite the SDI programme. SDI has been troublesome for nuclear arms control, but it was not the consequence of a reassessement of new technological possibilities emerging form military R&D. It was the product of visions and political ideologies. The base of ABM R&D that preceded SDI certainly provided that programme with a head start. But we should also note that the knowledge obtained from experiments, analysis and experience with ABM provided a better understanding of the technical limits of defence systems - making the case **against** SDI all the more credible!

Technological Innovation in Strategic Weapons

From an arms control perspective, not every substantial technological change in weapons systems was negative. Many advances in nuclear weapons technology resulted in improved performance of weapons but in ways that did not intensify the arms race (or at least should not have done so). Examples of military technologies considered even beneficial are devices that added to the safe handling and controllability of nuclear explosives (i.e. permissive action links and insensitive high explosives), systems that provide early warning of missile attack (i.e. infra-red sensing satellites) and systems that provide the essential national technical means of verification of arms agreements (i.e. reconnaissance satellites). For the most part, technological advances incorporated into succeeding generations of strategic systems since the beginning of the missile age (with the notable exception of accurate MIRVs) can hardly be considered alarming in and of themselves.

Looking at US systems, starting with ICBMs, advances in propellant technology and in techniques for hardening missile silos permitted succeeding generations of ICBMs to progress from the first generation of large, above-ground, cryogenically-fuelled missiles to small, solid-fuelled missiles based in hardened missile silos. The current generation of US silo-based ICBMs can presumably withstand overpressures perhaps 50 times greater than those of the very first silo-based missiles, and is immeasurably less vulnerable to a first strike than the very first above ground ICBMS that the USA deployed.

Soon after the start of ICBM development, new technology permitted ballistic missiles to be based in submarines as well as on land. Further advances in technology allowed for succeeding generations of submarine-based ballistic missiles to have increased ranges, thereby vastly increasing the submarine operating area. Also, as a result of technological efforts, submarines have become much quieter. In addition, basic investigations in the potential of non-acoustic techniques for ASW revealed that there were no breakthroughs likely very soon. With these developments, the threat of ballistic missile submarines becoming vulnerable has essentially been eliminated, at least for the forseeable future - a vital factor in strategic stability.

In the case of the long-range bomber force, the issue is a bit more complex, since vulnerability was always real by virtue of the existence of a very substantial, continuously modernized, Soviet air defence system. The choice for the USA, was either to exploit new technology to maintain the bomber force as a central element in its nuclear deterrent, or to eliminate the force and rely solely on ballistic missiles. Most analyses suggest that maintaining a long range bomber force has helped in efforts to achieve strategic stability.

But, of course, strategic weapons modernization has at times taken other, far less benign, directions. Particularly alarming was the development of MIRVs, which were originally justified as the ultimate ABM penetrators. To the extent that this discouraged ABM deployment and a consequent offense/ defence weapons race, it was good. But as we all know, MIRVs combined with high-accuracy delivery added to perceptions of vulnerability and in that way were a major stimulus to the arms race. While we publicly espoused MIRV as the "good", anti-ABM device, we were always aware and exploitive of the counterforce capability of MIRV.

It is interesting to note here that despite the obvious problem MIRV deployment on both sides would create, and despite the verifiability of a ban on full-scale testing and deployment, neither side seriously pursued the possibility of banning MIRVs prior to any deployments. The reasons seem clear and demonstrate some of the problems of trying to limit military technology. MIRVs had a very substantial immediate military utility - at little cost they vastly increased the number of targets that could be destroyed. In the narrow sense, MIRVs were very cost-effective. The USA was not about to give up the advantage it enjoyed of a substantial lead in this technology and wanted to proceed as rapidly as possible to deployment; the Soviet Union was not about to freeze its MIRV lag and wanted to complete its development and testing, to be on top of this technology, before agreeing to any deployment bans. So we see here that neither being ahead nor being behind in any technology assures interest in limiting that technology.

The modernization of the US strategic forces required a great infusion of new technology and was extremely costly. In retrospect, the pace of strategic weapons modernization was hardly warranted by the military threat as it actually evolved. But the political leaders who approved every major new system seemed to be influenced primarily by the political environment, by a perceived need to convey strength and determination and to influence world perceptions, rather than by what the new technologies had to offer. It is not at all obvious that a slower rate of progress in military technology would have resulted in a substantially slower rate of weapons acquisition or in fewer strategic weapons than the unconscionable quantity we have or that our strategic posture would be any more stable than it is.

Concluding Comments

1. Arms control has not been very successful so far in stemming the growth of, and preoccupation with, military weapons. There is natural interest

in trying to impose limits on the very early stages of weapons technology - military R&D. But in trying to keep weapons from reaching readiness for deployment we must realize that this may be feasible and effective only in well-defined cases and only at the stage of full-scale major components and system testing.

2. The main emphasis in arms control should remain on limiting or totally banning production and deployment of different classes of weapons. In the case where total bans on deployment are considered, bans or severe limits might also be applied on full-scale testing. This would provide added assurance that a deployment ban is being observed and that it would be difficult to develop a quick break-out capability from an agreement. It is the ban on weapons (be it on ASAT, space-based ABM, etc.) that is central; the limit on tests could only be a corollary to such a ban.

3. Negotiating broad limits on military R&D without much prospect for success is likely to raise issues, concerns and controversy about arms control generally and in that way would be counter-productive to arms control efforts. What we can strive for are self-imposed constraints by the application of "unilateral common sense" (a term coined by McGeorge Bundy) in considering what should and should not be pursued in military R&D.

Conversion from Military Research and Development: Economic Aspects

Judith Reppy

Introduction

One of the goals of arms control has been to reduce the economic burden of military spending by making possible the conversion of resources from military activities to civilian use. To date there is little evidence that this promise has been fulfilled, although the ABM Treaty of 1972 did allow the United States and the Soviet Union to avoid large expenditures on strategic defence systems for some years. But now there is hope that the more dramatic strategic arms reductions under negotiation could result in real savings - provided, of course, that they are not followed by intensified technological competition in strategic weapons or by a build-up of conventional arms.

Reaping the full benefit of reductions in military spending for the economy, however, will require conversion of the military resources to civilian activities. There must be jobs and retraining programmes available in the civilian sector for military personnel, and defence contractors must be able to produce and sell for civilian markets. In short, successful conversion will depend on a growing civilian economy and flexibility in redeploying resources to new uses. It will require a political commitment to overcome resistance from interest groups that benefit from military spending, but, in turn, the opportunity to convert to new activities can lessen that resistance.

Military research and development (R&D) occupies a special position in the discussion of the economic benefits of conversion. New technology is an important element in future economic performance, so that the opportunity cost of military R&D is not just the current value of the resources tied up in military activities, but also includes the implied loss of future civilian productivity gains that might have resulted from employing those R&D resources in civilian projects. Thus, even though military R&D is, in general, only a small part of total military spending, this reallocation to civilian uses could have substantial consequences for civilian economic performance. In particular, conversion could free scientists and engineers to work on those technologies, such as computers, microelectronics and new materials, that are currently transforming a large number of industries in the civilian sector.

Joint use of common technologies and spin-off of military technology to the civilian sector may offset in part the opportunity cost of military R&D. Since knowledge of new technology can often flow between the military and civilian sectors of the economy, the civilian economy can benefit from new military technology without paying the full cost of its development, although an investment in adapting the technology will be necessary. Some nations, in fact, consciously use military R&D programmes to bolster the technological level of their economies[1].

Historically, military-based technology has been important in civilian product developments such as commercial jet aircraft and nuclear power where the civilian and military functions are closely related, and in the military contribution of basic and applied research to a shared technology base. Unfortunately, from the point of view of the civilian economy, the vast bulk of military R&D spending is for engineering development of specific weapons systems; in the United States, for example, spending for the technology base programmes (excluding the SDI programme) is only 12 per cent of the budget for military RDT&E. Spin-off from the later stages of weapons development is relatively limited and is, in any case, a remarkably inefficient way of generating innovation for civilian use, compared to direct investment in civilian R&D projects.

Given the benefits to the economy that could accrue from conversion of military R&D, we must ask, what are the obstacles? Is there a conversion problem for military R&D in the same sense that there is a conversion problem for military spending in general? The resources engaged in military R&D are, after all, the most highly valued of all economic resources in both capitalist and socialist countries. At first glance it would seem that re-allocating skilled scientific and technical personnel to civilian activities would not be difficult, compared with the problems of retraining and re-locating the much larger number of workers engaged in military production and working at military bases.

Viewed from another angle, however, it is clear that there is a substantial problem. In the major countries performing military R&D - that is, in the United States, Soviet Union, Britain and France - military spending for R&D has been a very important part of total R&D activity throughout the post-World War II period. Currently, in the United States it is 70 per cent of federally-funded R&D and 30 per cent of total R&D spending. Western estimates of Soviet allocation of scientists and engineers place 50 per cent in the military sector[2]. The picture is only slightly less dramatic in Britain and France, and in those third world countries, like India, that maintain sizeable military R&D programmes. (Twenty-five per cent of a central government spending for R&D in India is for military programmes[3]).

In these countries there is a strong culture of military R&D, which at times has been seen as the most demanding and interesting work, as well as the best-funded. For example, Bruno Weinschel, President of the Institute of Electrical and Electronics Engineers (USA), is quoted as saying of defence work: "I had the finest lab facilities. State-of the art. It's fascinating - much more interesting than trying to save a couple of pennies manufacturing a radio"[4]. Conversion from military R&D will require altering those perceptions, as well as creating new opportunities in the civilian sector. It is likely to require more than a simple shift in funding priorities, which might well be seen as temporary and subject to reversal, as has so often happened in the past. Such a basic shift in attitudes and funding may only be possible in the context of a reduction in armaments generally as a result of reduced international tensions. Nevertheless, it is possible to analyse the conversion of military R&D as a separate issue, with its own benefits and difficulties, and to identify the sector-specific changes that it requires.

An Analytic Framework

The task of conversion is to shift resources from military to civilian uses. In centrally-planned economies this task is, in principle, a matter that can be decided by government fiat, although there may be internal political difficulties in reaching a decision to cut back on military spending. I focus here on the industrialized market economies, where conversion must take place through adjustments to market conditions that can be analysed in terms of supply and demand - in our case, the supply of and demand for the trained scientists and engineers who are the main resources employed in military R&D programmes.

Broadly speaking, demand for these resources is derived from governmental decisions about military programmes, on the one hand, and a more diffuse set of decisions about civilian development projects, on the other. Supply is a function of past investment in education and training, and the desirability of alternative employment opportunities. The market environment will differ between industries and countries and over time. Two elements are singled out for discussion here: technology and institutions. Together they go a long way towards defining the context in which governmental policy must operate in seeking to redirect R&D from military to civilian uses.

Technology

Military technology differs from civilian technology mainly in the characteristics of the end products. Some military products, such as submarines and guided missiles, have no civilian analogues, and, even in the case of products that are in common use, such as aircraft and trucks, the military design will differ from its civilian counterpart in order to meet military requirements.

At the components and sub-components level, there is more overlap in technology, although military requirements may still force a divergence in design for items that otherwise appear to be interchangeable. For example, specialized military requirements for testing, radiation hardening and ceramic packaging, distinguish military elcectronic chips from equivalent civilian chips and require special manufacturing processes.

From the perspective of conversion, the most significant differences between military and civilian technologies is the military emphasis on the development of large, costly, high-performance systems that are procured in small numbers, leading in turn to labour-intensive production processes. Scientists and engineers who have devoted their entire careers to working in these conditions may find it very difficult to adjust to another setting. It follows that civilian projects that resemble military products - i.e., large systems with a high R&D content - are plausible targets for expansion to absorb workers from the military sector. Civilian space projects, energy, especially nuclear energy, and transportation projects are obvious examples. Successful conversion depends at least as much on adjusting to commercial market conditions as to civilian technology requirements; to the extent that the market for these products resembles the military market with its single customer and limited competition, the transition will be easier.

At the other end of the spectrum, scientific manpower that is employed in developing generic technology and components for use in both military and cvilian products can be switched between uses with relative ease. The military market is likely to be a small subset of the total demand and the pay-off for new technology is in the commercial sector. Barriers to conversion at this level are not attributable to the characteristics of the technology, but rather to institutional rigidities, such as military specifications and secrecy.

Institutions

Opportunities and incentives to shift from military-related work will differ according to the institutional setting. First, and most obvious, the degree of central government control over the economy determines what policy measures are available and appropriate for redirecting resources. Centrally-planned economies, and countries with a well-established consensus in favour of a national industrial policy, have available a range of policy instruments that are not politically feasible in a country like the United States, in which the propriety of a large role for government in the economy is vigorously debated.

The nature of the military establishment and its enthusiasm for new technology, especially for large leaps in technology, are also important. Where new military technology is given a high priority and indigenous development of technology is favoured over imports, military establishments will have large organizations dedicated to research and development. These R&D organizations develop requirements for new weapons, oversee their development and perform some of the R&D in inhouse laboratories. Their **raison d'etre** is tied directly to the military mission, so that a shift away from military work within the organization is not likely; conversion will require the scientific workforce to shrink, either through job mobility or retirement. Related governmental laboratories outside the military hierarchy, such as the nuclear weapon laboratories in the United States, generally have civilian as well as military tasks. For them conversion can take place through the expansion of the civilian side, e.g., alternative energy projects, as the military projects are scaled back although there remains the problem of successfully transferring the technology developed in the laboratories to the private sector[5].

In market economies defence contractors, whether private companies or state-owned enterprises, perform most of the military R&D. For these companies, military R&D programmes are essential for future business because they lead to production contracts. The companies work closely with their military counterparts to develop requirements and to increase funding for military R&D to meet those requirements. Many of these enterprises sell in both civilian and defence markets, but their defence business tends to be segregated in a separate division of the company, with relatively little contact with the commercial divisions.

Conversion here is a matter of shifting workers across the dividing line into civilian projects, of finding civilian customers for products manufactured in the defence divisions, or reducing the size of the defence firms in favour of growth elsewhere. Success in commercial markets for defence contractors require fundamental changes in attitudes to cost, reliability, and marketing, a transformation that will not be easy. In the past, attempts of defence contractors to diversify into commercial markets have, with few exceptions, been notably unsuccessful[6].

For all these institutions new programmes of military R&D are vital for future organizational growth or even continued existence. Hence, they may be expected to resist any moves to reduce spending. Overcoming that resistance is a political task that must be accomplished, if policies for economic adjustment are to succeed. In a sense, the economic programmes for conversion suggested below may be thought of as bribes to the defence R&D community, designed to subsidize the private contractors in unfamiliar commercial markets and to entice the technical workforce into moving into civilian work.

Policies for Conversion

Conversion of military R&D will require strong effective demand in the civilian sector to provide markets for the increased civilian output produced by resources transferred from the military sector and, on the supply side, an adaptable and mobile labour force. To the extent that the required demand conditions are a by-product of macroeconomic policies for full employment and growth (or, in socialist countries, decisions of the central planning organization), there is little specific that need be said about them here. In the past, when economic growth has been strong - for example, in the years following World War II - the conversion of resources from military to civilian use has proceeded with relatively little difficulty. When macroeconomic conditions have been unfavourable, as in the early 1970s in the United States, reductions in spending for military R&D have led to unemployment of scientists and engineers[7].

Beyond general fiscal policies, there is scope for governmental policies targeted specifically towards attracting scientific and technological resources from the military to the civilian sector. In all countries funding for military R&D programmes comes from the state, which can choose to allocate its funds elsewhere. Spending for government programmes for civilian-based R&D, e.g., in the field of energy, can be increased as military spending is decreased. As noted above, these projects resemble the large-scale systems development of the military sector, and thus require less adjustment on the part of defence contractors than products designed for consumer markets. Dual-use technologies, such as new materials or fibre optics, can be promoted under the rubric of civilian rather than military programmes. And there are other governmental policies, such as tax incentives, that can be introduced to stimulate the demand for civilian R&D.

Conditions affecting the supply of scientific resources are important also. Here the question is whether labour shifts easily or is slow to move in response to changes in the market for R&D. The flexibility and mobility of the scientific and technically trained labour force will depend on a number of factors and may well differ between countries. In general, when demand for scientists and engineers is growing, there will be a response both from newly trained entrants to the labour pool and from lateral movement out of related fields. For example, over half the increase in the number of astronomers that occurred in reponse to the growing US space programme in the 1960s was met by physicists and others moving into the field from other jobs[8].

The response of the labour supply in the military sector when demand is decreasing - the conversion scenario - is more problematic. The difficulty or ease of shifting to new jobs will depend on the education, experience and adaptability of the technical workforce employed in military R&D and the degree of overlap between the military projects and civilian projects that may replace them. The overlap between military and civilian communications satellites, for example, is very high, and it has been common for engineers to move back and forth between the two kinds of projects[9]. But other areas of military technology - for example, missile design - lack civilian analogues. Workers who have devoted a whole career to working in these specialized and often highly secret fields may find it very difficult to adjust to another setting.

Other factors affecting the elasticity of supply are whether the workers have had opportunities for continued education and training in changing technical fields and the age profile of the workforce. Presumably, technical personnel who have had a broad-based education and training, have not spent many years in military work, and have participated in continuing education, will find it easier to shift out of military work into civilian projects. Unfortunately, the United States, at least, is characterized by an aging work force in the defence industry, and these older workers may find adjustment to employment in civilian projects particularly difficult. In this case the solution may lie in gradual attrition through retirement, rather than shifting to a new career.

The dominant culture of the defence sector, with its emphasis on technical bravura, may handicap defence workers who move to the civilian sector. Accustomed to designing for performance with little regard for cost they may not be able to adjust to civilian industry, which must compete on the basis of the price and reliability of its products. Again, basic engineering training that emphasizes flexibility and adaptability could reduce the significance of this problem. Policies to increase competition in military markets, such as those pursued in recent years in the United States and Britain, may succeed in bringing defence contractors closer to the civilian practice. Similarly, a shift in defence procurement policy to emphasize use of civilian technology and components wherever possible, could reduce the size of the defence-dependent industry.

Conclusions

Redirecting scientific resources from military to civilian projects requires both a shift in demand and a response on the supply side. The demand condition can be met through a combination of macroeconomic policies and industrial policy targeted to civilian technology-intensive industries. An increase in emphasis on dual-use technology for military requirements could, in itself, cause a shift in the locus of R&D activities to the civilian sector, and direct promotion of civilian projects would create additional alternative employment opportunities.

On the supply side, shrinking the size of the military R&D-performing sector will require a long-term commitment to educational and training programmes to increase lifetime flexibility in the technical workforce. A practical first step might be changes in the curriculum of engineering schools and company policies to emphasize broad basic training with opportunities for retraining at periodic intervals; in an age of rapid technological change these reforms would generally benefit workers and companies.

For conversion measures to be successful they must be part of a fundamental reordering of priorities. Otherwise, contractors will be reluctant to invest in the development of civilian projects, scientists and engineers will resist redirecting their careers, and the political interest groups will maintain pressure to restore military programmes. Fortunately, a virtuous circle is possible, and over time the restructuring and reduction of the military R&D sector should lead to a reduction in the political and organizational pressures that sustain it.

References

1. J. Molero, "Foreign Technology and Local Innovations: Some Lessons Learned from Spanish Defense Industry Experience", in The Relations Between Defence and Civil Technologies, P. Gummett and Judith Reppy (eds.), Dordrecht: Kluwer, forthcoming, 1988).

2. David Holloway, The Soviet Union and the Arms Race, 2nd ed., New Haven: Yale University Press, 1984, p. 134.

3. F.A. Long, "Science, Technology and Industrial Development in India", in Framework for Interaction: Technical structures in Selected Countries Outside the European Community, H. Fusfeld (ed.), Center for Science and Technology Policy, R.P.I., New York, 1987, p. 11 and 13.

4. Quoted in Rosy Nimroody, "Star Wars: The Economic Fallout", Council on Economic Priorities Research Report, CEP Publication N87-11, Nov/Dec 1987, p. 3.

5. "Labs Struggle to Promote Spin-offs", Science, 240, 13 May 1988, pp. 874-76.

6. M.L. Weidenbaum, "Industrial Adjustments to Military Expenditure Shifts and Cutbacks", pp. 253-87 in The Economic Consequences of Reduced Military Spending, B. Udis (ed.), Lexington, MA.: Lexington Books, 1973, and Economic Adjustment/Conversion, Report prepared by the President's Economic Adjustment Committee and the Office of Economic Adjustment, OASD (MI&L), Washington, D.C., 1985, ch. 7.

7. National Research Council, The Impact of Defense Spending on Nondefense Engineering Labor Markets, Washington, D.C., National Academy Press, 1986, p. 3.

8. E. Ginzberg, J.W. Kuhn, J. Schnee and B. Yavitz, Economic Impact of Large Public Programs: The NASA Experience (Salt Lake City: Olympus Publishing Co., 1976, pp. 150-51.

9. Presentation by J. MacLucas at the Harvard Conference on Dual-Use Technology, Boston, MA., 22-23 April 1988.

Part VII

Global Environmental Problems

The Environmental Dimensions of National Security

Paul Ehrlich and Anne Ehrlich

Climatic Effects of Nuclear War

While the attention of national leaders in both East and West is focused mainly on the balance of their conventional and nuclear military forces - "star wars", and military turmoil in the Middle East and Central America - they are allowing the very basis of national security to evaporate. Politicians, from habit and ignorance, view the political and economic differences between American and Soviet societies as of paramount importance. In an historical context, those differences are no greater than those, say, between Romans and barbarians fifteen centuries ago. Yet some American and Soviet politicians apparently are willing to blow up the world over their differences. Fortunately, Roman emperors did not have access to nuclear arms in 376-476 A.D. as the barbarians overran the empire. If they had, the Soviet Union and the United States probably would never have existed.

That the geopolitical spheres of interest of today's two superpowers have little actual overlap is rarely noted. Equally, too few of the world's leaders realize that the rapid depletion of Earth's non-renewable resources, the deterioration of the global environment, and the related widening of the economic gap between nations of the industrialized North and the developing South represent an unparalled threat to their nations' security. That threat not only portends a continual deterioration of living standards virtually everywhere in time of peace, but also contributes to conflict between nations[1] and thus increase the chances of nuclear war.

A large-scale nuclear war between the East and West blocs could destroy civilization in the Northern Hemisphere in a very short time. Prompt deaths could number 500 million - one-tenth of the world's people - or more. The World Health Organization[2], using a scenario that was perhaps too pessimistic, estimated that over 1000 million deaths could result directly from an all-out nuclear war. Subsequent environmental effects mediated primarily through atmospheric changes, could multiply the number of deaths severalfold[3,4].

There has been controversy over the magnitude of climatic changes that would follow a large-scale nuclear conflict - that is, the severity of a "nuclear winter". But results of sophisticated recent atmospheric studies[5,6] indicate conditions that would cause extremely severe damage to biological systems even in the most optimistic models.

It would make little difference to corn or wheat, for example, whether the sky was pitch black and temperatures fell to -40°C in July in Iowa, Bordeaux, and Kazakhstan, or whether there merely were a few weeks of twilight, chilly weather, and scattered frosts. Whatever agricultural potential remained in the absence of inputs and social organization would be destroyed in either case by fluctuating temperatures, darkness, frosts, high levels of ionizing radiation, acid precipitation, increased UV-B flux, and other less widely discussed atmospheric sequelae of the war. Civilization clearly could "end with a bang".

Disruption of the Natural Ecosystem

Destruction from war is not, however, the only catastrophic threat to security the world faces as it moves towards the twenty-first century. A fundamental source of both the increasing danger of conflict and the NorthSouth economic gap is the finite nature of Earth. The quantities of resources that humanity can mobilize are limited by the planet's size and physical characteristics. So is the capacity of Earth's environmental systems to absorb the ever-increasing damage inflicted upon them as a consequence of human activities - including mobilization of resources. The more gradual undermining of security as global environmental problems worsen is arguably just as serious as the consequences of a nuclear war but much less widely perceived.

Growth of the human population is a major factor in that erosion of security. Today there are over five billion human beings, and one and a half million are being added each week. The world is increasingly overpopulated, and the symptoms of it are becoming more and more evident. The clearest evidence of overpopulation is that today's five billion can only be supported (with a billion or more living in deepest poverty) by consuming a one-time inheritance from Earth[7,8].

Humanity is destroying its environmental capital - not just petroleum and other mineral resources but rich agricultural soils, "fossil" groundwater, and biotic diversity (populations and species of other organisms). These resources are not renewable on the time-scale of interest to society. They have accumulated over periods ranging from millennia to hundreds of millions of years, but they are being dispersed and destroyed in decades to centuries.

So even if no third world war occurs, current trends in human population growth and **per-capita** impact on the global environment may well destroy civilization. The main difference from the consequences of a nuclear war would be that, rather than collapsing in a matter of months, it would take decades. Rapid climate change, insidious soil erosion, intractable pollution problems such as acid deposition, and a general decline in ecosystem services such as flood and pest control, would almost certainly set the stage for a gradual collapse by reducing agricultural productivity. And epidemic diseases, terrorism, ethnic conflicts, revolutions, and "limited" regional wars could easily play substantial contributing roles. Civilization thus could "end with a whimper".

The key point here is that security has many dimensions other than the military one. While Earth's nations altogether spend roughly a trillion dollars annually on armed forces, the overall security of every country is rapidly diminishing.

Ending civilization with a whimper would amount to a gigantic "tragedy of the commons"[9]. People and nations acting independently in their short-term best interest create situations that in the long term destroy the common resource they are all using. As Garrett Hardin wrote[9], "Ruin is the destination toward which all men rush, each pursuing his own best interest in a society that believes in the freedom of the commons. Freedom in a commons brings ruin to all." His example was overgrazing of a village's common pasture; civilization's ruin will stem from treatment of the biosphere as a commons with almost no international controls and the organization of human societies in ways that make it difficult to recognize or act on that behaviour.

Perhaps no feature of Earth is more a commons, a feature of the planet used by all and owned by none, than the atmosphere. The atmosphere connects all nations and people, past and present, and it interacts intimately with oceans and land, including plants, animals, and soils, as a major component of ecosystems.

A central factor in the erosion of global security today is disruption of the functioning of Earth's natural ecosystems as the scale of human activities increases. Ecosystems are a complex of the plants, animals, and microorganisms of an area, interacting with each other and with their physical environment. Most people take ecosystems for granted and are unaware of their importance in supporting human life. Natural ecosystems provide an array of essential services, without which civilization would soon collapse and which technology cannot replace. In some cases, scientists would not know how to begin replacing nature's services; in others, technological substitutes are inadequate and far too costly.

Several ecosystem services centrally involve the atmosphere. Ecosystems help keep the air breathable, ameliorate the climate, and operate the hydrological cycle that supplies fresh water. The atmosphere also participates in the delivery of other services to the degree that its qualities help support the lives of the organisms providing the services. Forests are key to flood control, but if the atmosphere continues to be loaded with oxides of nitrogen and sulphur, producing acid precipitation, that ecosystem service may ultimately be lost. Vegetative cover is essential for the preservation and replenishment of soil as well as retention of moisture. Loss of vegetation often leads to drought; drought in turn can kill remaining plants, leading in turn to denuded surfaces, erosion, and (by changing the land surface's albedo) further climate change.

Natural ecosystems dispose of the wastes of civilization and in the process recycle nutrients essential to themselves and to agriculture. They supply food from the oceans and freshwater systems, provide pollination and pest control services to crops, and maintain nature's vast "genetic library". The last service is especially important. Humanity has already withdrawn from that library the very basis of civilization in the forms of domestic plants and animals, innumerable industrial products, and medicines. Yet, the potential of the library has barely been tapped.

Links with Security

The condition of the atmosphere is especially closely linked to security. Because it is a global commons, shared by all nations to an even greater degree than the oceans, the atmosphere's health is a vital concern to everyone; yet no one is responsible for maintaining its integrity. Yet, human-generated changes in the atmosphere could lead to problems between nations - indeed, they have already begun to do so.

Over the next few decades, an increase of a few degrees in average global temperature is expected to follow the build-up in the atmosphere of greenhouse gases such as carbon dioxide and methane. This build-up is traceable to various human activities, especially the burning of fossil fuels, deforestations, increased paddy-rice cultivation, and expansion of cattle herds. A warming of the atmosphere will alter patterns of atmospheric circulation, changing both the climate and the composition of the atmosphere (e.g., its load of dust or pollutants) over large areas.

Consider some of the more obvious possible effects of such an occurrence. Summers in northern mid-latitudes could become hotter and drier. The agricultural productivity of the great plains of North America might be severely reduced by heat stress and less dependable rainfall. While the now marginal climate of much of the Soviet Union's wheat belt might be improved in the long run by greenhouse warming, the process of adjustment could further destabilize that nation's shaky grain production system over the short term.

The economic situations of both nations could deteriorate seriously, and the USA might be unwilling or unable to sell grain abroad. If the Soviets were in need of imported grain, tension between the superpowers could rise, and those tensions might be aggravated by agriculutral problems in other regions, which could no longer obtain food imports. Over 100 nations now depend to one degree or another on grain imported from North America to feed their populations[10]. The disappearance of surplus grain production in the United States, which could result from a combination of climate change and depletion of ice-age water in aquifers beneath the great plains, could lead to a global crisis.

Altered rainfall patterns, which are virtually certain to accompany a global atmospheric warming, could also exacerbate tensions over increasingly scarce water supplies in places as disparate as the Middle East and southwestern North America. Egypt is already tottering on the brink of chaos, and a further reduction of the Nile's flow could push it over the edge. If rainfall became sparser in Israel and its neighbours, an explosive situation would become much worse. The waters of the Jordan, Yarmuk, and Litani rivers have been sources of tension for years; and Israel, which consumes five times as much water **per capita** as its neighbours, may have a serious water shortage in the 1990s[11]. The hardnosed Israeli stand on the West Bank and Gaza in 1988 may be just a foretaste of what is to come.

International tensions also could be heightened by reduced river flows between Pakistan and India (over the Indus) and India and Bangladesh (Ganges). Disputes over water have occurred in those regions and in the basins of the Tigris/Euphrates, Amur, Mekong, Rio de La Plata, and Nile rivers. The water of 120 of the world's 200 major river systems is shared by two or more countries[12]. With the entire planet overpopulated and serious regional water shortages already widespread, the potential for drought-induced conflict is enormous.

For instance, a significant decline in the dependable supply of water available from the Colorado river could lead to further trouble between Mexico and the United States. Such a conflict could be greatly intensified by more Mexicans seeking to enter the United States in order to make a living. The flow of ecological refugees, already serious in many parts of the world, seems bound to increase. Agricultural conditions in many regions are deteriorating as productivity declines because of soil erosion, failing irrigation systems, desertification, or change in climate.

In some cases, as suggested earlier, a shift in climate may be benign and might help to relax international tensions. But even where changes are beneficial in the long run, disruption arising from the change itself could create problems, especially if it occurs rapidly[13]. Increased rainfall in semi-desert areas thus might manifest itself primarily as flooding, depressing food production and bringing downstream nations into conflict with upstream ones that open dams and pass floodwaters downstream.

Transnational air pollution problems also will be affected by global warming, but in ways that are virtually impossible to predict. Forest systems, already stressed by air pollution and acid precipitation, are not likely to respond well to abrupt changes in temperature and rainfall regimes. Indeed, there is evidence that some forests in the northeastern United States are already being degraded by the combined stresses of pollutants and climatic change[14].

In the longer term, atmosphere-ocean coupling and changed freshwater flows could affect the dynamics of oceanic fish populations and, temporarily at least, depress productivity. Global **per-capita** fisheries harvests have been declining for almost two decades, a trend that is likely to continue even without changes in oceanic environments that are (in terms of evolutionary time) very rapid. Human population growth will not slow down enough to allow fisheries production to catch up. Indeed, coastal pollution and the destruction of estuaries (which serve as critical nurseries for many marine species) could soon lead to declines in absolute (rather than **per-capita**) production. Shots have been fired over fishing rights; that may only be a hint of things to come unless the portion of the Law of the Sea covering management of the fisheries commons can be extended and enforced.

Effects of Climate Change

Climate change, deforestation, and widespread pollution obviously will seriously interfere with maintaining the crucial "genetic library" function of natural ecosystems. Undisturbed forests, grasslands, or other natural areas throughout the world increasingly are islands of habitat in seas of human disturbance. The vast majority of land area on Earth has been taken over for human use and more or less drastically altered in its biotic character.

More than a tenth of the world's land is used to grow crops; perhaps two per cent is paved over or covered by cities and towns; a quarter serves as pasture for livestock; and most of the 30 per cent that is forested is exploited at some level or has been converted to tree farms. Most of the remaining 30 per cent of the planet's land is in arctic or antarctic regions, in desert, too mountainous, or otherwise too inhospitable to be of much use to civilization.

Reflecting this pervasive takeover of the planet, recent calculations indicate that humanity directly or indirectly accounts for a large and growing fraction of Earth's biological productivity - the energy produced by photosynthesizing organisms (such as green plants, algae, and many kinds of bacteria) using sunlight - on which virtually all animals and other non-photosynthesizing organisms ultimately depend. The total energy annually captured worldwide through photosynthesis, and not used by the photosynthesizing organisms to run their own life processes, is known as net primary production (NPP). Human beings and domestic animals directly consume (as food, fodder, and timber) about four per cent of the NPP produced on land and about two per cent of that in the oceans[15].

Although this seems modest (if a disproportionate share for only one of perhaps 30 million species of animals), the indirect human impact on productivity on land is far from modest. Including the diversion of much productivity into human-directed systems, and therefore into supporting different communities of organisms (such as agricultural pests) than in natural systems, multiplies the human impact on terrestrial NPP to over 30 per cent. Included in this calculation are the NPP produced in recently converted pastures, material burned on grazing lands, plant materials killed though not used in timber harvesting or in land-clearing for agriculture.

Furthermore, the indirect impacts on terrestrial NPP extend beyond the large fraction that is co-opted in human systems. The conversion from natural to human-dominated ecosystems more often than not results in reduced productivity. Cropland (except when irrigated) is usually less productive than the grassland or forest it replaced; pasture is less productive than a forest. In addition, people have reduced productivity in many areas by paving them over and through desertification - usually a result of over-grazing, over-cultivation, or poorly managed irrigation. Since the middle of this century, it is estimated that global terrestrial NPP has been reduced by roughly 13 per cent[15].

Humanity's direct consumption, indirect co-option, and suppression of photosynthetic production thus add up to an impact of nearly 40 per cent of the planet's NPP on land, and a total global average (including the lesser impacts on oceanic systems) of about 25 per cent. This enormous diversion of the energy source of all life on Earth goes a long way in explaining what is happening to natural ecosystems, why so much of terrestrial life is endangered, why ecosystem services seem to be deteriorating in so many parts of the world, and why, as a consequence, all nations are less secure.

The fraction of land worldwide that has been set aside specifically to preserve natural ecosystems is a pitiful three per cent. With the exceptions of some polar regions and the swiftly vanishing moist tropical forests of South America, Southeast Asia, and parts of Africa, little land today remains outside those reserves that is still both biotically valuable and unspoiled enough to qualify as a reserve. And, especially in developing nations, with their rapidly growing human populations, nature reserves are vulnerable to encroachment and degradation. The world's poorest and fastest growing populations are mostly in the tropics; there also nature's diversity is most bountiful. By some estimates, much more than half of all the species on Earth are found in tropical forests, which today occupy only about seven per cent of the land surface and are shrinking rapidly in the face of human expansion.

Paradoxically, the biotic wealth in tropical forests is usually not convertible to agricultural wealth - at least not through conventional modern agriculture. Removal of luxuriant tropical vegetation more often than not leads to poverty and desertification. As ecosystem services are disrupted, the local climate changes, topsoil and nutrients are quickly lost, and pests are unleashed to devastate crops[16].

To compensate for these problems, which are greatly accentuated in the nonseasonal tropics, higher levels of inputs (fertilizers, pesticides, etc.) must be supplied (and their environmental side-effects tolerated) for a relatively meagre return in crops. Results of cattle or other livestock production in tropical regions are similar: small returns, often only for a few years, after the effort of land-clearing. A frequent outcome of continuous cropping or pasturing is a few years of production followed by abandonment of the degraded land and clearing of another patch of forest. In such circumstances, as deforestation proceeds and local ecosystem services decline, the region's agricultural production is anything but secure.

The process of forest clearing for cropland and pastures poses a clear threat to nature reserves, which are increasingly isolated as islands of forest surrounded by farms or an abandoned wasteland. In some countries, the isolated reserves may be essentially the only undisturbed land left. The reserves thus become vulnerable to invasion by land-hungry farmers and ranchers, or by people cutting trees for fuelwood or poaching the protected animals. Reserves degraded by these activities have a diminished capacity to provide ecosystem services, already reduced by impoverishment of the surrounding land.

Even if nature reserves could be flawlessly guarded against human invasion, wood cutting, and development, they cannot be protected against climate change and air pollution. Terrestrial organisms have evolved in response to climatic regimes. Most populations are relatively intolerant of substantial and rapid changes in those regimes. Throughout Earth's history, species have responded to climate change by gradually changing their geographic distributions and by evolving tolerances to new conditions.

But today migration is often not possible, since it would have to occur through impassable barriers of disturbance. And the rate of climate change now envisioned for the near future will be much too rapid for populations of all but the fastest reproducing organisms (such as bacteria and some insects) to have a chance of surviving through genetic adaptation. Thus, the integrity of the remaining fragments of natural ecosystems will be further undermined by changing climate in addition to the more direct assaults from human activities - and the ecosystem services essential to the support of the human enterprise (especially agriculture) will be further destabilized.

Climatic change of course is not always unfavourable, as noted above. Even in the short run, some agricultural areas with marginal rainfall now might receive increased precipitation and produce larger or more stable crop yields. In the long run, warmer temperatures or increased precipitation might well open new areas to agriculture, although problems with soils would have to be overcome in many regions. But overall, one must assume that rapid climate change will cause more harm than good, simply because social systems, agricultural systems, and natural ecosystems are adapted to current conditions and tend to change slowly. Just the process of adjusting to an unpredictable new climatic regime will be disruptive enough to shake even the most secure agricultural economies.

Other large-scale atmospheric problems facing humanity include increased UV-B radiation as a consequence of stratospheric ozone depletion (ozone depletion, of course, is also involved in climate change), acid deposition, and other forms of air pollution. But serious though these are, climate change has a greater potential for contributing to international conflict and for diminishing the economic security of all nations. Aside from largescale war, no single facet of the human predicatment is more frightening.

In A.E. Housman's worlds, we must "train for ill and not for good". We cannot risk the future of civilization on the hope that climate change might improve the lot of humanity as it struggles in the next few decades with the gravest problems it has ever faced. The world is burdened now with overpopulation, depletion of resources and environmental deterioration. Humanity is suffering from an increasing maldistribution of wealth, including the growing gap between rich and poor nations, rampant racism, sexism, and religious prejudice - all of which add fuel to the fires of competition and conflict.

As if these were not enough, deterioration of the human epidemiological environment seems likely to supply another source of international tension. That environment becomes ever more precarious as human numbers grow, average susceptibility to disease increases (with malnutrition on the rise), and the speed with which disease carriers can move around the globe also increases. Malaria is resurgent, and AIDS has the potential for destroying a large portion of humanity. The AIDS virus is the third lethal virus to threaten a large-scale epidemic since 1967.

Whether or not AIDS can be contained will depend primarily on how rapidly education can slow its spread, when and if the medical community can find satifactory preventatives or treatments, and, above all, luck. A virus that infects millions of people may well evolve new transmission characteristics. For instance, if the ability of AIDS to be transmitted heterosexually increases or it evolves the ability to concentrate in saliva, an already disastrous situation could worsen. Fringe groups have demanded that AIDS victims be quarantined, and blood tests are now required of immigrants into some nations. If the epidemic continues to spread, disruption of trade and international recriminations are real possibilities.

The Need of Worldwide Co-operation

There is, however, one positive aspect to the dangers presented by the AIDS epidemic, the prospect of climate change, acid precipitation, and other large-scale environmental problems: these clearly are issues that must be dealt with cooperatively by the international community.

In the modern world, with the increasing dependence of all nations on world trade and communications networks, avoidance of serious problems originating elsewhere is less and less possible. Surviving a nuclear world war or a pandemic without consequences therefore is unlikely for most nations, if only because the resultant social disruption elsewhere would have severe repercussions on trade, communications, and possibly migration. The seeds of conflict contained in transboundary environmental problems such as pollution, acid rain, and export of hazardous substances should be of concern to all nations, whether directly involved or not. Similarly, resource depletion can produce serious economic dislocations far from the main scene of action, as the OPEC petroleum cartel clearly demonstrated in the 1970s.

The increasingly tightly-knit world society sooner or later will have to come to grips with these and other regional and global problems, if only because they generate tensions between nations and could lead to war. But one global problem is universal and inescapable: there is no practical way for any segment of humanity to avoid the consequences of a rapid change in global climate. No nation's policies by themselves could assure national security in an era of worldwide forest death, desertification, widespread crop failures, and general ecological collapse.

The rate of climatic change must be slowed, its effects minimized, and global environmental deterioration reversed if nations are to find anything resembling security. To do so will require worldwide cooperation. In essence, the post-world-war II movement towards world government must be revitalized and strengthened. Success in establishing international regulation has historically been achieved by bringing it in through the back door, by creating agencies to regulate in areas in which national governments effectively did not function - such as the Law of the Sea. The latter succeeded in large part by granting power and responsibility to individual nations to manage resources in their adjacent waters.

Surrendering national sovereignty has always been the main stumbling block to establishing a world government. But far more international regulation and management of commons now take place than most people realize. Most of this is accomplished through a complex maze of trade, political, monetary, and other kinds of agreements and through the activities of the United Nations' many agencies and some other independent ones such as the World Bank. In addition, much is done, if not always for universal benefit, by private organizations from multinational corporations and the global stockmarket network to non-government organizations such as Planned Parenthood, Oxfam, and CARE.

In effect, a world government does exist, but it is radically different in structure from most national governments, principally lacking central administrative or legislative bodies and therefore unrecognized. Administrative functions are carried out through a diffuse system of agencies which usually have no real power over nations, although a kind of international peer pressure often works. A World Court also exists, but it too has no power to enforce its decisions. Unfortunately, the two superpowers, whose observance of decisions would most influence other nations to comply, are among the worst scofflaws.

A Global Commons Regime

The population-environmental-resource-economic super-problems, that loom in our future for the next generation or two, demand a stronger, better recognized, and respected system for managing the global commons. It may not be possible - or desirable - to create a strong centralized world government patterned after national governments, but a great deal of scope exists to build on the present model of diffuse, semi-independent agencies with different (though often overlapping) responsibilities, which tend to solve problems through consensus and international agreements.

Thus, for instance, building on the Law of the Sea idea, a Global Commons Regime might be established to regulate human interactions with the atmosphere and oceans. A little reflection shows, of course, that such a regime would have to regulate to some degree such diverse activities as agriculture, the clearing of tropical forests, the generation of power, the use of international combustion engines, the manufacture of chemicals, plastics, and hair sprays, and the disposal of toxic wastes.

There is no guarantee that an effective Global Commons Regime could pull humanity through the coming crises; but it seems certain that, without such an effort to deal with global environmental problems, civilization will collapse. In the face of the common threat to all nations and peoples, the political quarrels now expressed in minor wars and revolutions, and even the competition between the East and West blocs, pale by comparision. One could hope that the overwhelming need to collaborate in averting the worst consequences of these super-problems could impel nations to forge new alliances and find avenues of cooperation - to find peace. What needs to be done will require an effort that is nothing less than monumental. But the benefits for all of humanity if the atmosphere/ocean commons could be cooperatively monitored and protected would also be monumental.

A central problem facing us now is finding ways to convince national and international leaders and the world's people that opportunities for action to rebuild global security are fast slipping away. A number of actions to slow the rate of climatic change can and should be taken now by individual nations, and international agreements to reduce competitive disadvantages in trade could follow quickly.

Among the actions possible now are adoption of policies to increase the efficiency of fossil-fuel and wood energy use, and promote development of energy-mobilizing technologies that do not deposit CO_2 in the atmosphere, including the development of nuclear power reactors that have safety as a primary characteristic. To reduce emissions of methane (another greenhouse gas), cattle herds could be reduced and development of alternative food sources to replace beef encouraged. The manufacture of chlorofluorocarbons and other chemicals that cause destruction of stratospheric ozone should be phased out as quickly as possible; for essential uses of such chemicals, the search for substitutes should be accelerated.

One more essential measure to decrease the atmospheric buildup of CO_2 - indeed, to reverse the process as forest growth absorbs CO_2 - is to subsidize the preservation and (where necessary and feasible) restoration of natural ecosystems in poor nations. Indeed, restoration of vegetative cover and regeneration of forests everywhere should be near the top of the human agenda. This move is needed for other important reasons besides slowing the build up of CO_2; halting and reversing the deadly spread of deserts, slowing the horrendous global rate of agricultural soil erosion and protecting agricultural productivity, and providing source of fuelwood for poor people are among the other benefits that would be realized. In short, some of these measures would contribute directly to enhanced security and well-being.

The pioneering Guanacaste project in Costa Rica, regrowing tropical seasonal forests, could provide a model for similar projects in the developing world. Another might be completion of the proposed joint USA-USSR restoration and preservation project for Madagascar's forests.

Overpopulation

Above all, humane programmes to bring population growth to a halt and begin a slow decline are desperately needed. Overpopulation both contributes directly to the factors causing global super-problems, such as climatic change, and makes their potential consequences more dire. Civilization must move as rapidly as possible towards a number of people that Earth can support in reasonable comfort primarily **on income**[7,8], The prospects of trying to deal with the dislocations caused by climatic change alone (to say nothing of the other elements of the human predicament) in a world inhabitated by eight to ten billion people are daunting indeed.

Consider, for example, what would happen if there were a birth control miracle in India and that nation achieved replacement reproduction around 2025. In that case, India's population would continue to grow until almost the end of the next century, and when it stopped India would have about two billion people - the number on the entire planet when we were born in the early 1930s. Picture what monsoon failures would mean to two billion Indians! Imagine what might happen if they shared the Indian subcontinent with close to 300 million Pakistanis and both nations were well armed with thermonuclear weapons. It would not be a situation conducive to international tranquility.

Recently the **Club of Earth**, a group of eleven scientists who are both members of the US National Academy of Sciences and fellows of the American Academy of Sciences, prepared a statement on population. It begins:

> "Arresting global population growth should be second in importance on humanity's agenda only to avoiding nuclear war. Overpopulation and rapid population growth are intimately connected with most aspects of the current human predicament, including rapid depletion of non-renewable resources, deterioration of the environment, and increasing international tensions."

Conclusion

Since, regardless of any actions started now, environmental problems will get vastly worse as Earth becomes even more overpopulated, planning should be started now to address intelligently and humanely the inevitable dislocations. First, part of the money now being wasted on weapons should be diverted into enterprises that will increase the security of nations.

Establishing many more agricultural experiment stations and extension services worldwide to help farmers improve their food production and respond to climate change is an obvious requirement. So is improvement of long-term food storage and transport facilities to help insure against catastrophic famines. We can all think of others. The actions needed to help humanity survive the crisis in the decades ahead are fairly obvious; generating the social and political will to accomplish them will be horrendously difficult.

References

1. Westing, A.H. (ed.), 1986, Global Resources and International Conflict, Oxford University Press, New York.

2. WHO, 1984, "Effects of Nuclear War on Health and Health Services", Geneva.

3. Ehrlich, P.R. 1988, "The ecology of nuclear war", in P. Ehrlich and J. Holdren (eds.), The Cassandra Conference, Texas A & M Press, College Station, pp. 181-99.

4. Harwell, M.A., Hutchinson, T.C., **et al.** 1985, Environmental Consequences of Nuclear War: Ecological and Agricultural Effects, SCOPE, 28, Vol.II, New York: John Wiley & Sons.

5. Covey, C., Thompson, S.L. and Schneider, S.H. 1985, "Nuclear winter": a diagnosis of atmospheric general circulation model simulations, J. Geophys. Res. 90, pp. 5615-28.

6. Schneider, S.H. and Thompson, S.L. 1988, "Simulating the climatic effects of nuclear war", Nature, in press.

7. Ehrlich P.R. and Ehrlich A.H., 1986, "World population crisis", Bull. Atomic Sci., April, pp. 13-19.

8. Ehrlich, A.H. and Ehrlich P.R. 1987, Earth, New York: Franklin Watts.

9. Hardin, G. 1968, "The tragedy of the commons", Science 162, pp. 1243-48.

10. FAO, 1986, Production Yearbook, 1985, Rome.

11. Cooley, J.K. 1984, "The war over water", Foreign Policy 54, pp. 3-26.

12. Myers, N. 1985, "The environmental dimensions of security issues", Mimeo.

13. Schneider, S.H. 1988, "The greenhouse effect: What we can or should do about it", Proc. First North Amer. Conf. on Preparing for Climate Change, in press.
14. Hamburg, S.P. and Cogbill, C.V. 1988, "Historical decline of red spruce populations and climatic warming", Nature 331, pp. 428-30.

15. Vitousek, P.M, Ehrlich, P.R., Ehrlich, A.H. and Matson, P.A. 1986, "Human appropriation of the products of photosynthesis", Bio Science 36:6, pp. 368-73.
36:6, pp. 368-73.

16. Ehrlich, A.H. 1988, "Development and Agriculture", in P.R. Ehrlich, and J.P. Holdren (eds.), The Cassandra Conference, Texas A & M Press, College Station, pp. 75-100.

Environmental Problems: A Determining Factor of Future Politics

Erhard Keppler

Introduction

Three major topics are dealt with in this paper: the climate problem; the acidifiers introduced into the ecosystem; and the health-risks which industrial societies suffer due to increasing emissions into the atmosphere, rivers and soil. There is some relationship between these, since carbonburning also creates SO_2, NO_x as well as CO_2. Release of chloro-fluorhydrocarbons into the atmosphere destroys stratospehric ozone; these gases in particular seem to be responsible for the Antarctic ozone-hole, which might reduce the global ozone inventory further, so that skin-cancer increase may become a real threat. As these other gases are also relevant to the greenhouse-effect, their increasing concentration in the atmosphere links their emissions also to the climate threat.

Acid Deposition and Health

It is now fairly well established that the deposition of SO_2 and NO_x - either as "acid rain" or in dry deposition - has developed into a serious hazard to living plants, to animals, and to men. SO_2 and NO_x are mainly produced by carbon-burning, besides some natural sources, like volcanoes, or N_2O release from soils (which becomes partly oxidized in the troposphere). It is likely, that this increased SO_2 burden of the air, in particular in cities, is responsible for the increasing rate of skin-damage and attacks on man's respiratory organs, including pseudo-croup and cancer. In the presence of carbon hydrates and NO_x during sunshine periods, ozone and other superoxides are being produced in concentrations which are close - and in densely populated areas often beyond - the human toxicity threshold; surface cells of organic material (plants, animals) are also damaged and thus contribute to what is called "Waldsterben" in various ways, including synergistic effects. This damage depends strongly on local situations, and therefore varies in extensity as well as in appearance.

However, the acidifiers of the strong inorganic acids also reach the soil. As pH-values are lowered, typical chemical reactions in soil release metal-ions, including finally iron, which eventually enter aquifer systems.

This mobilization of metal-ions results in decreasing fertility of soils, but has of course also severe effects on plants (including wood), depending on the specific composition of the soil. Alkalinity has been shown to become the biological relevant factor, apart from the pH-value. The continuous release of these inorganic acidifiers is, therefore, not only a pertinent factor to forest devastation but affects also agricultural production. In many areas the yearly amount of lime which has to be brought to agricultural areas exceeds 2-3 tonnes per hectare; all this just to keep the pH-value above 6, which is necessary to guarantee some yield.

Natural sources of SO_2 are volcanoes and wood fires. SO_2 is also generated in the atmosphere from hydrogen sulphide through oxidation by the hydroxyl-radical OH. Estimates are varying, but a crude figure of volcanic emissions is 30 million tonnes of sulphur per year, while production from H_2S is 90 million tonnes per year. This compares with anthropogenic emissions of 100 million tonnes of sulphur per year, of which 90 per cent is related to energy production. This figure has not improved (despite strong efforts in some countries to remove SO_2 from emissions) mainly due to increasing sulphur emissions in many other countries (burning of coal and oil). About 93 per cent is emitted in the northern hemisphere. NO_x is generated in natural processes, in soils and by lightning, at a rate of 20 million tonnes of nitrogen per year, at about the same rate from wood fires and tropospheric NH_3-oxydation, and a similar quantity from anthropogenic coal-burning.

A potentially severe threat is the ongoing progression of the souring-front in soils, which moves down towards the aquifer system. Mobile heavy metal-ions increase their concentration in ground water, which is the major supply for drinking water. Heavy metal-ions are not harmful to men as long as their concentration does not exceed certain levels. By continuous addition, concentrations will sooner or later reach these critical levels; drinking-water would then become poisonous to men, if not chemically treated prior to use.

Climate

Climate is a long-term feature of our atmosphere. It is the result of complex physical and chemical processes, among which transport (horizontal and vertical) controlled by the solar energy input is the most prominent. The sensitivity of our climate to changes in such processes results from the fact, that it is in a "dynamic equilibrium", which means that the system is in radiation balance. However, thermodynamically it is in disequilibrium. Climate is thus the mean global long-term temperature pattern. For any constituent of the atmosphere there are source- and loss-processes; its concentration is determined by the balance between them, as is its mean life-time. Among loss processes are mechanical, vertical and/or horizontal, motions of airmasses, and chemical reactions - like the oxydation of hydrogen-sulphide to carbon dioxide. The life-time determines whether a particular constituent becomes globally mixed, or whether local gradients exist.

Solar heating affects the temperature variation with height, and also thermal- and pressure-gradients between low and high latitudes. It, therefore, affects chemical reaction rates, and source-and loss-processes of the various atmospheric constituents. This complicted feedback mechanism makes atmospheric physics one of the most complicated mathematical problems that have ever been attacked by man. Up to now only approximate solutions to partial problems have been solved. It must be said, however, that due to the introduction of very sensitive and new measuring techniques (which improved the database in a significant manner), and due to much faster and larger computers and new computational methods, sensational progress has been made recently in modelling climate responses to variations in atmospheric composition, which affect the radiation budget.

A major effect in the global energy budget is the greenhouse effect. This effect is caused by multi-atomic molecules in the atmosphere, which due to their atomic constitution are able to absorb energy in the infra-red wavelength domain. They reabsorb a fraction of the energy, which the earth tries to radiate into cold space in order to maintain its energy budget. If such molecules absorb radiation, they contribute to a temperature-increase in their environment, which eventually averages over larger regions, thus warming the biosphere.

Such gases are water vapour and carbon dioxide, but also ozone, nitrogenoxides, the CFCs (discussed below), methane, and other multi-atomic molecules present in the troposphere. It is worthwhile to mention, that molecules like ozone, present in the stratosphere in concentrations of a few per million other air molecules, remove almost all of the solar ultraviolet radiation preventing it from reaching the earth surface. It is, therefore, important to recognize that mixing ratios of even less than one part per million are significant in respect to climatic effects. The important issue is that increasing emission rates of such gases occur in the atmosphere, and we cannot ignore their climatic effect any longer[1].

Spurious Gases Relevant to Climate

We cannot control the water content of the troposphere; it varies according to weather conditions, and may reach a few per cent. On a global scale, the degree of cloudiness is about constant, so that the global energy budget due to water vapour variations is balanced on a daily basis. For this reason we can ignore water vapour and focus on the other gases. If the earth had no atmosphere its, surface would have had a temperature of -18°C. Since the global average temperature is +15°C, the greenhouse effect amounts to 33°, of which about 20° are due to water vapour present in the atmosphere. The rest is due to other gases as shown in Table 1; which also gives their rates.

Table 1

Abundance of Some Spurious Gases in the Earth's Atmosphere

Gas	Rel. Abundance (parts per million)	Rate of Increase (per cent per year)	Contribution to Greenhouse Effect (per cent)
CO_2	338	0.5	46
CH_4	1.5	1-2	9.7
N_2O	0.5	1.3	6.5
O_3	0.02-0.05	~ 0	9.7
CO	0.05	1.2	< 1
CFC	< 0.003	up to 15	28.1

Many gases have natural sources; methane, for example, is generated by a broad variety of decay processes in soils and swamps; ammonia is produced by certain bacteria, laughing gas stems from microbiological denitrification in soils. During the last hundred years of industrialization a growing amount of gases from industrial processes was, and still is, released into the atmosphere. The atmosphere has for long been considered as an infinite storage; the idea that even this sink may become over-loaded became accepted only during recent years. Waste release into the atmosphere has always been considered to be something like a basic human right - today we recognize, that use of the atmosphere, like the use of rivers and soil as waste-deposits, must be paid for. We have learned that global warming will result from such releases.

Today most dangerous anthropogenic gas releases into the atmosphere are the emissions of carbonoxides, of methane, and of halogenized carbonhydrates (referred to as CFCs).

Carbon dioxide

CO_2 is generated by burning (oxidation) of carbon (coal, oil, gas, wood). Such oxydation also occurs as a consequence of deforestation in tropical areas, where light may disintegrate the fragile organic acids in biomasses, formerly protected by the forests, thus releasing CO_2. It is also generated by agriculture, in cement production, in volcanic eruptions - and of course by breathing of living biota. The carbon cycle removes CO_2 from the atmosphere during assimilation, the carbon becoming fixed in organic material. Such material finally transforms into oil, gas and coal. CO_2 is also exchanged between air and oceans, and a complicated cycle is going on, whereby carbon dioxide is dissolved in water, as well as being transformed into bicarbonate (HCO_3^-) and carbonate (CO_3^{2-}) in a sensitive equilibrium system, which influences the transport into the deep sea and the gaseous releases back into the atmosphere. While the deep sea cycle operates at time constants of the order 1000 years, the cycle time of carbon in the upper reservoirs is of the order 10 years. It is, therefore, clear that the rate of increase of CO_2 in the atmosphere is essentially determined by the capacity and uptake rates of the ocean and the biosphere.

Present rates of CO_2 release from anthropogenic sources are continuously increasing with a tendency for even further acceleration[2] (Fig. 1). All CO_2 sources produce about 8 billion tonnes of carbon per year, leaving about 50 per cent in the atmosphere. About 50 per cent of the carbon-

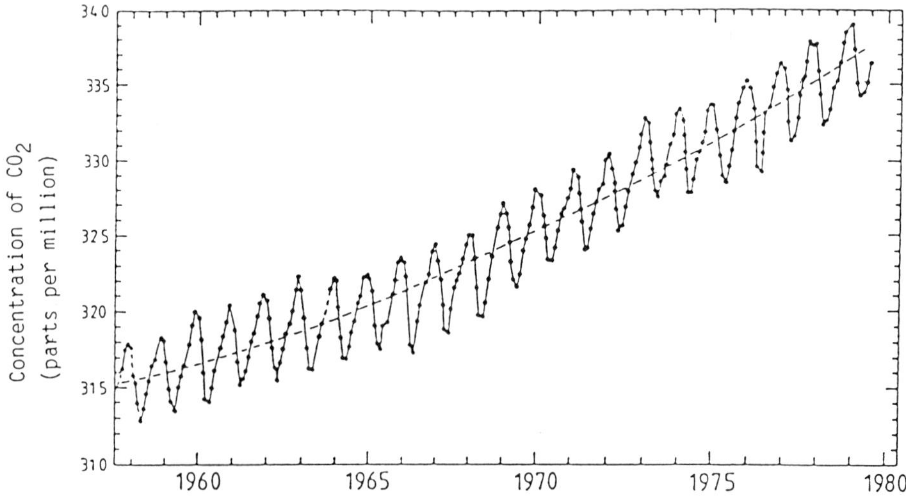

Figure 1 Increase of CO_2 with superimposed semi-annual variations due to assimilation, as recorded in Mauna Loa, Hawaii.

hydrates, produced during daytime assimilation, feed the respiration of the plants: there is a net removal of CO_2 from the atmosphere by green plants. A recent increase of green plant areas in northern higher latitudes is, however, overcompensated by rapid deforestation in tropical areas.

CO_2 absorbs light in bands between 4.5 and 15 μm within the earth's infra-red radiation window. It thus contributes significantly to the greenhouse effect and is most relevant for the climate problem.

Methane

Methane is generated under aerobic conditions by bacteria in swamps and soils, in rice-fields, in significant amounts by fermentation of ruminants (20 per cent!), the largest contribution apart from termites. Ice cores from Greenland and Antarctica give a very good record of the historic atmospheric methane contents. Fig. 2 shows its atmospheric concentration: fairly constant levels in the 16th century but an accelerated increase ever since[3]. It can be shown that this increase is directly proportional to the number of persons living on earth (the link being food production, i.e. rice-fields and ruminants.

Part of methane is destroyed in the troposphere by oxydation with the OH-radical, but a significant amount reaches the stratosphere; its flux is comparable to the upgoing water-vapour flux. In the stratosphere methane plays some role in the formation of hydrochloric acid, which is a temporary sink for free chlorine, and thus reduces the ozone destruction rate of chlorine in the stratosphere. In the troposphere, however, methane plays an important role in tropospheric ozone production: due to the removal of OH, it indirectly controls whether ozone is produced (in the presence of enough nitrogen oxide) or destroyed (NO-deficit, HO_2-production). This process plays an important role in the way plants are damaged by ground-level ozone and acid effects.

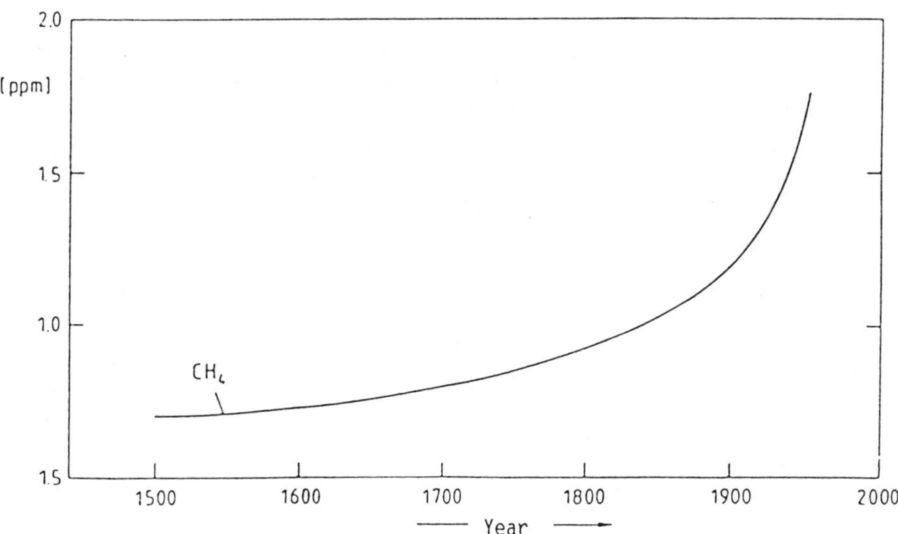

Figure 2 Increase of methane in the atmosphere

Methane absorbs the planetary infrared emission in bands between 8 and 8 μm, thus contributing to the greenhouse effect. The methane concentration grows at a rate of 1-2 per cent per year; its present contribution to the greenhouse effect is of the order of 10 per cent, it is, therefore, one of the gases, which must be watched closely in the future.

CFC

The chloro-fluor-carbonhydrates became the most threatening gaseous constituent of the earth atmosphere in recent years. Their effect on the ozone budget in the stratosphere has first been recognized in 1973. These gases do not react with other gases or material as they traverse the troposphere, transported upwards by turbulent mixing processes. In the stratosphere, however, they become photo-dissociated by energetic ultraviolet photons. The released chlorine and other halogens destroy ozone, and by further reaction with atomic oxygen become reestablished, ready for another cycle. This "catalytic" process allows one halogen atom to cycle some 100 000 times before it is eventually removed, e.g., as a hydrochloric acid molecule. It is this multiple cycling, which allows substantial ozone-destruction rates, despite the fact that their concentration is only in the one part per billion range.

Similar catalytic processes include, e.g., nitrogen oxides, which have strong natural sources. Their presence always contributed to the ozone losses, but this was recognized only recently. Because methane may reduce its effectiveness, the CFC-effect has been overestimated and had to be reduced several fold when better data and better models became available. This caused frustrating debates with chemical industries on whether CFC emissions should be reduced or even forbidden.

These processes are important because 30 per cent of the total atmospheric ozone is concentrated in heights above 25 km. Today we know that there is a net reduction in the ozone concentration in the 40 km height region of the order of a few per cent per year. At the same time an increase in the ozone concentration in the 10 km height region is observed. For this reason the total amount of ozone in a column remains essentially unchanged.

The growth rate of these gases, however, is incredible: ranging from 5 to 7 per cent per year. For a while the replacement of fully halogenated carbonhydrates by only partially halogenated gases (which contain hydrogen atoms and can be attacked by the hydroxyl-radical on their way through the troposphere) was thought to be a solution, as it was believed that the gases would not reach stratospheric levels. In the meantime, however, measurements have shown that they have reached concentrations comparable to those of fully halogenized CFCs. The growth rates have been shown to reach a frightening 15 per cent per year![4,5].

We, therefore, must expect to observe increasing ozone destruction rates in the stratosphere in future. These gases have lifetimes in the atmosphere which in some cases exceed 100 years, and are thus uniformly mixed over the globe, including Antarctica; their dominating role in ozone destruction in southern spring is becoming generally accepted. At present it is an open question, whether the ozone losses over Antarctica can be compensated for at all, or whether we have to expect further global ozone reduction due to the ozone hole. If so, such a reduction would certainly increase the ground level ultraviolet intensity. The threat of increasing appearance of skin cancer could then become reality.

Ozone reduction and a change in its height distribution is also an effect relevant to climate - as it changes the height distribution of the

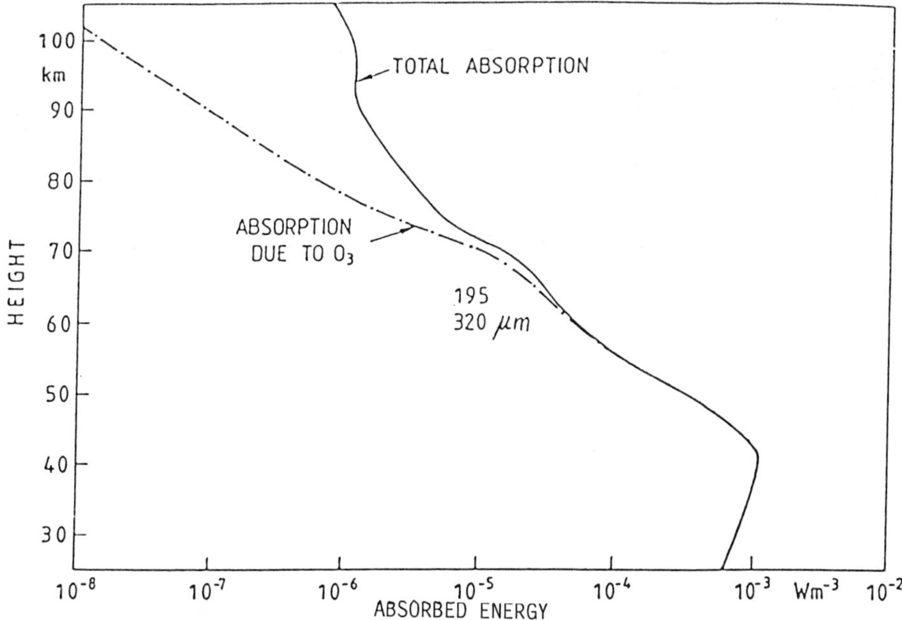

Figure 3 Differential energy absorption of solar radiation by ozone as a function of height, compared to contribution of other gases.

solar radiation, ozone is responsible for the largest differential energy deposition in the atmosphere (Fig. 3) and thus determines the atmospheric temperature profile to a wide extent. Any change will certainly influence the global atmospheric meridional circulation, thus also horizontal temperature- and pressure-gradients, and the climate.

CFCs like all multi-atomic gases have absorption bands within the infra-red thermal emission band of the earth (Fig. 4), and therefore contribute to the greenhouse effect. A conservative estimate gives a contribution of 30 per cent to this effect at present.

The Greenhouse Threat

CO_2, which increases due to a global annual increase of 3-4 per cent in consumption of fossil fuels, will reach doubling of its present mixing ratio in the middle of the next century. Such doubling will reduce the infra-red emission of the planet by 4.4 W/m^2, which will be further increased by feedback mechanisms. This will correspond to an increase in the average global temperature between 2° and 3°, but this may be delayed due to the heat capacity of the oceans.

The presence of the other gases with their much larger growth rates will, however, result in a temperature increase in a much shorter time. Doubling of the present mixing ratio will occur within 35 years at an annual growth rate of 2 per cent, but in only 14 years at a 5 per cent rate. The contribution of methane and CFCs will therefore reach the CO_2 contribution within 20 to 30 years from now, and thus shift the 2-3° temperature increase

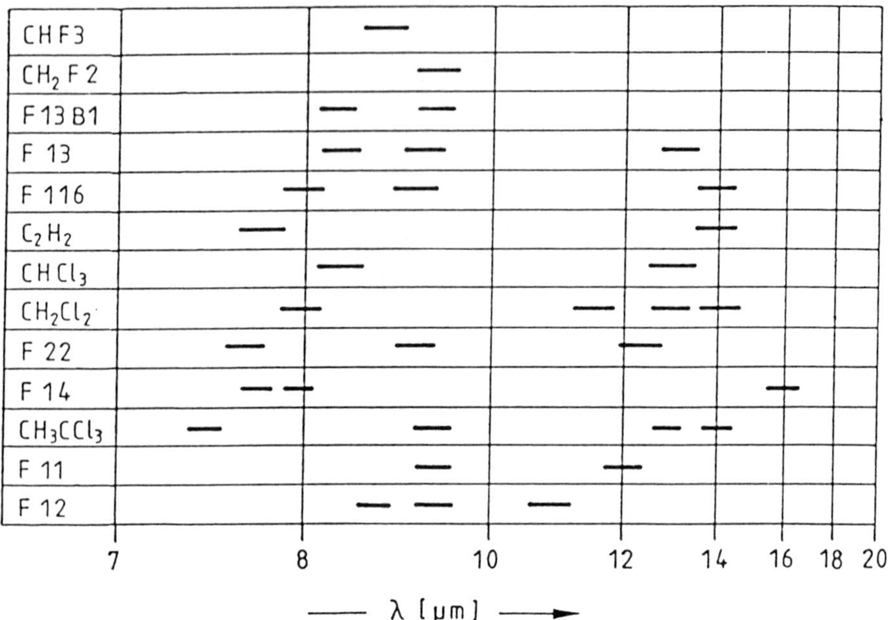

Figure 4 Absorption bands of various chloro-fluor-carbon-gases within the thermal emission band of the earth (3-50 μm).

to the first decades of the next century - if the present growth rates remain unchanged.

A further increase of the global average temperature by 2° will result in a temperature increase of about half that value in latitudes below 30°, but in twice that figure in latitudes above 60°. Model calculations, which include realistic cloud formation processes, point to increased cloudiness in high latitudes, accompanied by increased rainfalls, and reduced cloudiness in lower latitudes, which means further increase in desertification with all its consequences to the people living in such areas.

A first indication of the reality of these predictions is already seen in measured data (Fig. 5). In higher latitudes melting of ice in the Arctic sea will accelerate, followed by melting of the polar ice caps. This, and further melting of mid-latitude glaciers, will result in a continuous rise of the sea level, becoming gradually evident by an increasing number, and more heavy floods, followed by losses of coastal areas near sea level.

Predictions as to what the fate of a particular region will be cannot be made because orographic conditions, with their wind regimes, will greatly influence regional conditions, resulting in better conditions in some areas, but in catastrophic events in others. Almost certainly climatic changes will affect the distribution of the world's water supply, with enormous consequences for food and energy production, This means that certain areas will have to be abandoned, with people moving into other regions, i.e. a modern migration of nations must be expected. Recent experience with refugees and related problems gives some hint of what may be expected.

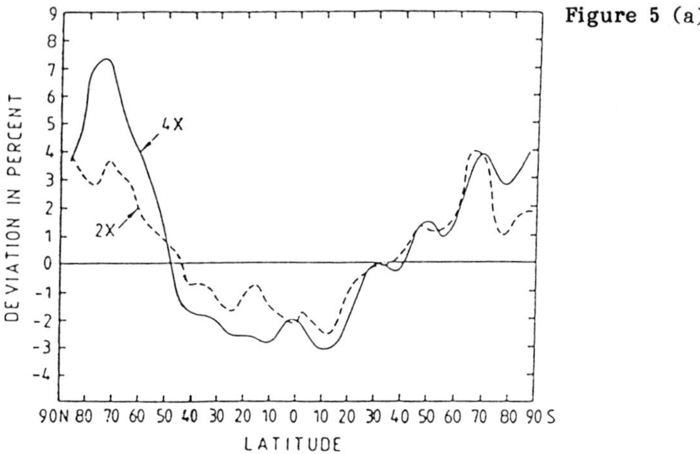

Figure 5 (a)

Model calculation (3-dimensional, including cloud formation processes) of total cloudiness as a function of latitude for two levels of equivalent CO_2 concentration[6].

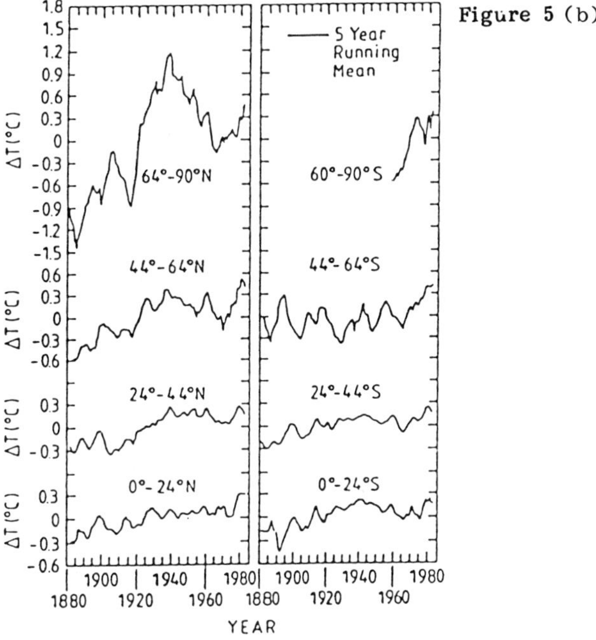

Figure 5 (b)

Measurements of surface air temperature increases in low, middle and high latitudes (5 year running means)[7].

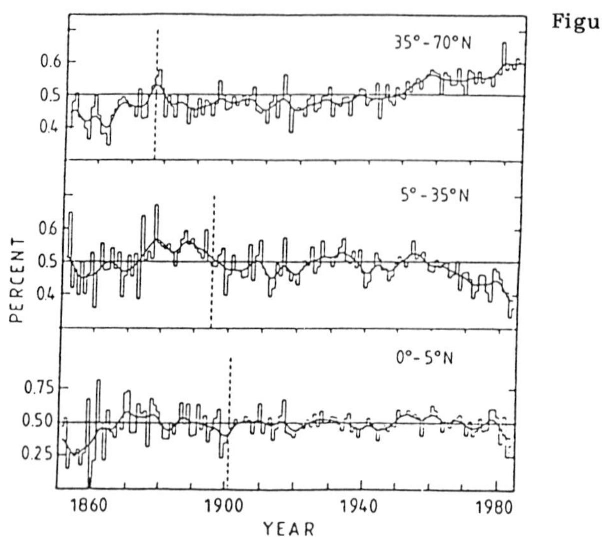

Figure 5 (c)

Measured precipitation index for low, middle and high latitude regions[8]

Physical Aspects

The earth's atmosphere is physically an open system that can maintain a thermodynamical equilibrium only by permanent energy supply from the sun. The atmosphere is illuminated by the sun, exchanges matter, energy and momentum with the cryosphere, the hydrosphere, the biosphere and the litosphere.

Such systems can spontaneously form spatio-temporal patterns[9] on macroscopic scales via self-organization, i.e. without any specific interference from outside. Turbulence is such a feature; the global atmospheric circulation pattern is another, quasi-permanent feature. Others are formation and disappearance of low pressure systems, hurricanes, etc.

Today we are able to describe the atmosphere on a microscopic level quite accurately. On a mesoscopic level we consider the atmosphere as a gas or a fluid, described by mesoscopic parameters like local density, temperature, pressure, mixing ratio of molecules. Here we deal with patterns as mentioned: deterministic chaos. Such patterns tend to form within characteristic timescales. Many of them could become catastrophic events. Whether we have time for protection or even counter-measures depends on the timescales. The birth of a hurricane may be recognized in time to warn people in affected areas; a climate change has much longer timescales, which will allow us to start counter-measures - if we recognize the signals, and let the precious time not just pass away.

What Should be Done?

We, members of the species **Homo Sapiens**, could, knowing the threat, quickly understand that the following worldwide measures should be taken:

1. We should reduce carbon-burning as much as possible and turn to renewable energy-sources, like solar, wind, or geothermal energy, and also to nuclear energy. Most important, however, would be energy saving - wherever energy is converted, efficiency must (and can) be improved. We should also start to reduce greatly carbon-fuel consumption in traffic, either by promoting public transport services, or by supporting electric driven vehicles. Air traffic should also be greatly reduced.

2. We should reduce energy demands by improving heat insulation and substitution of materials which require large energy input for their production.

3. We should immediately stop deforestation in tropical areas.

4. We should introduce strict birth control, thus stopping the increase of population, and eventually turning it into a decrease. An annual increase of 95 million people is beyond what the planet can support.

5. We should stop, worldwide, the production of CFCs.

6. We should set up a programme aiming at "closed" processes - i.e., processes, which neither emit gases nor fluids, and establish close control of waste deposition.

7. We should greatly reduce the use of fertilizers in agricultural productions, stop use of pesticides, use plant protection chemicals only after proven compatibility with life on earth. Such agricultural production can probably support only a limited population - smaller than the present one.

Nobody is likely to deny that such goals are desirable, but whether they may be achieved remains questionable. On the other hand, I feel it is necessary to have such goals in order to adjust planning, even if they may be reached only partly and in steps, spread in time over some years. But I still hope that we have a good chance to preserve living conditions on our planet.

A Change in View is Required

The Earth is the only planet in the solar system which preserves life. Life may have originated on Mars as well, also in other stellar environments - which we do not know, but probably not on Venus. Even on Mars we have not yet seen life, or traces of life that may have existed there in the past.

On Earth life has now existed for almost 3.7 billion years. Nevertheless, the Earth is not distinguished within the solar system, nor is the solar system. Nature cannot be considered to exist "for man". Modern natural sciences do not give any support to such views. The universe obeys the physical laws, as far as we know, without any exception. The still increasing greenhouse effect, which mankind has set going on Earth within a century, can be understood in physical terms, its future development - known production rates for the relevant gases - can be calculated. Such calculations have been found to be in general agreement with fundamental observations.

The conditions which human life can tolerate are also well known: temperature range, atmospheric pressure, electromagnetic radiation (spectral band and intensity therein), ionizing radiation level, gravity, composition of gases around us, in particular the oxygen contents, and chemical composition of our food. Whenever any one of these boundary conditions are violated, man would not survive. History has very many examples to prove this.

Life had always taken the chance to adapt itself to a changing environment, but time scales for evolutionary modifications are in the million years range. Ongoing evolution may bring about life forms which will be adapted to an environment that would exist at the end of this large geophysical experiment, mankind is presently performing. Nature is bound to rise and decay - also human beings are only transient participants in this ongoing process. Nature has not been made for man as was widely believed since the renaissance period. **Homo sapiens** had his chance - he should not wantonly risk it.

We are in danger from parasites, which adjust much faster to changing environments; we are threatened by acid gas relases, by chemical reactions in the atmosphere with direct (e.g. photo-oxydation) or indirect (ozone destruction with subsequent UV-increase) effects, by increasing amount of waste (gases, fluids, solid material), and finally by changing the atmosphere's composition with subsequent climate changes. It is justified to say that mankind behaved up to now in a reckless manner. In view of the consequences, it must be stated that future political measures must be based on scientific insight, with proper judgement of the effects which any single decision may have.

What We Must Do - An Outlook

I do not believe, that we are left with no chance, even though there is almost no indication that reasonable joint measures will be taken in the near future. Climatic changes must not occur in big jumps (even if in the past such jumps did occur): they must be smooth. We will probably start to observe calamities like the drought in the Sahel-zone more often, and, in parameters relevant to climate, we will see an increasing frequency of deviation from the mean.

This gives us, concerned scientists, the chance to inform people, politicians in particular, convince them that measures must be taken in a short time. We have to make these problems known, get them acknowledged within nations. This should lead to regional improvements - within the European Community, North America, the Socialist Countries, South-East Asia, etc. At the same time the United Nations should take the lead with respect to international conventions regarding air cleanliness, stop the increase in population, reduce coal burning and gas emissions, control waste and so on.

We must reduce worldwide emission rates of gases relevant to the climate less than 30 per cent of the present concentration. This corresponds to the reduction of such emission by 2 per cent per year. If we postponed such reduction measures by 20 years, we would then have to reduce emissions by 7 per cent per year. While the former can be realized, it seems doubtful whether the latter may be reached - given the present economic situation. We would then really be in deep trouble. This has also been clearly stated by the German Physical and Meteorological Societies[10].

We may have a decade to reach first results, and two decades to really have the above mentioned goals within reach. This gives us a realistic chance. But we must start **now** to slow down emission rates, so that we gain time.

Of course, the heaviest burden for such measures falls on the countries which produce most of the greenhouse gases, i.e. industrial nations, USA, USSR, European Community, Japan, China. Population growth, on the other hand, has its major contributions in threshold and developing countries (but also China and USSR must be addressed). As social and religious influences in some countries often oppose rational insight, control of birth will probably not come about quickly. Droughts and other catastrophic events will very likely substitute human inability and reduce populations. Deforestation of tropical areas will be stopped only if the industrial countries stop their demands for these woods; on the other hand, they must agree to pay higher prices for agricultural products from these countries, as well as seeking an acceptable solution to the debt problem.

Only after these old questions are solved, will politicians attack the global, new problems mentioned above. We must expect that some nations will, due to geographic advantages, not agree to establish effective birth control, or realize energy-saving programmes, or reduce coal burning. We must, therefore, urge that in international negotiations such behaviour is punished by trade restrictions and embargos. This is much more justified than present-day decisions based on moral grounds. After the turn of the century we will see that we are talking about bare survival, not about ethical judgement of some groups. Counter-measures against those countries who obstruct necessary actions will then be justified on objective grounds.

In doing so we will recognize that we have to deal seriously with the "North-South" problem; we will also recognize that negotiations on disarmament, though important, will become minor problems, and East-West differences may no longer be considered to be of prime importance. I am convinced, that in the years to come there will be one major topic to be dealt with: preservation of the planet Earth, our only base to live and to survive.

References

1. E. Keppler, Die Luft, in der wir leben, München: Piper-Verlag, 1988.
2. USDOE, Summary of the carbon dioxide effects, Research and Assessment Program, Washington, D.C.: US-Department of Energy, 1979.
3. M.A.K. Khalil, and R.A. Rasmussen, "Secular trends of atmospheric methane", Chemosphere, 11, p. 877, 1982.
4. P. Fabian, R. Borchers, B.C. Krüger, and S. Lal, "The vertical distribution of $CHClF_2$ (CFC-22) in the stratosphere", Geophys. Res. Letters, 12, 1, 1985.
5. P. Fabian, "Halogenated hydrocarbons in the atmosphere", Handbook of Environmental Chemistry, Vol.4/Part A, Berlin-Heidelberg, Springer-Verlag, 1986.
6. E. Röckner, "Simulation der Wolkenbedeckung in einem Klimamodell", Ann. Meteorologie (NF), 23, p. 90, 1986.
7. J. Hansen, and S. Lebedeff, "Global surface air temperature: Update through 1987", Geophys. Res. Letters, 15, 323, 1988.
8. R.S. Bradley, H.F. Diaz, J.K. Eischeid, P.D. Jones, P.M. Kelly, and C.M. Goodess, "Precipitation fluctuations over Northern Hemisphere land areas since the mid 19th cenutry", Science, bf 237, 171, 1987.
9. H. Haken, (ed.), Evolution of order and chaos, Berlin-Heidelberg-New York, Springer Verlag, 1982.
10. Deutsche Physikalische Gesellschaft e.V., Deutsche Meteorologische Gesellschaft e.V., Aufruf: "Warnung vor drohenden weltweiten Klimaänderungen durch den Menschen", 1987.

Global Environmental Issues and International Politics

Peter Gleick

Introduction

Over the last few decades, concern over the international implications of large-scale environmental problems has begun to command the attention of policymakers. The landmark 1972 United Nations Conference on the Human Environment in Stockholm discussed the right to a healthy and productive environment, including adequate food, safer water, and clean air. Since that time, some progress has been made, notably in improving air and water quality in industrialized nations. But enormous problems remain to be solved and environmental deterioration in developing countries continues at an alarming rate.

At the same time, a variety of new complex threats have arisen, including acid precipitation, Arctic haze, ozone depletion, species extinction, and global warming. Unlike previous concerns about air and water pollution, these threats have implications that are truly global in scope - they know no political distinctions and they obey no international boundaries. Global environmental problems thus threaten to upset international relations, behaviour, and security.

The issue of global climatic change - the so-called "greenhouse effect" - epitomizes these concerns. With the exception of nuclear war, no other environmental problem has the scope or the potential for such widespread societal impacts. Yet until recently, the concern about greenhouse warming has been the domain of the natural scientists, rather than the social or political scientists. As the technical consensus about the nature and severity of the problem has grown however, policymakers have finally begun to express concern.

> "Environmental threats to security are now beginning to emerge on a global scale. The most worrisome of these stem from the possible consequences of global warming caused by the atmospheric build-up of carbon dioxide and other gases... Slowing, or adapting to, global warming is becoming an essential task to reduce the risks of conflict[1]."

The role of global environmental problems in affecting international relations is discussed here, with a focus on global climatic changes. Particular attention is paid to the different consequences for developed and developing countries and the role that the industrialized and developing world may play in preventing or mitigating the worst societal impacts.

Environment and International Security

A wide spectrum of non-military threats should lead to a major reappraisal and expansion of the traditional, narrow concept of national or international security.

"...defining national security merely (or even primarily) in military terms conveys a profoundly false image of reality. That false image is doubly misleading and therefore doubly dangerous. First, it causes states to concentrate on military threats and to ignore other and perhaps even more harmful dangers. Thus it reduces their total security. And second, it contributes to a pervasive militarization of international relations that in the long run can only increase global insecurity[2]".

Environmental stresses are both a consequence and a cause of political tension and international disputes[3,4]. Many factors affect the connections between environmental stress, poverty, and security, including inappropriate development policies, resource constraints, population pressures, and economic inequities. Although these links are complex and poorly understood, it is widely acknowledged that resource constraints or environmental problems can lead to conflict when other pressures and tensions exist between states. In this way, resources act as **triggers** (e.g. Middle East oil), as **tools** (e.g. weapons links to energy, financing, accesss to materials), as **consequences** (e.g. targeting of dams, nuclear reactors, or oil fields), and as **roots** (e.g. economic competition, pressures, and tensions) to conflict.

Among all the major environmental threats, global climatic changes are most likely to affect international politics and security because of their wide scope and magnitude. Climatic changes may cause either a direct shortage of a resource such as water or food, or they may lead to the degradation of resources held either in common or exclusively. In the first case, disputes and conflicts may arise by competition for a limited resource, by the perception that competition may arise for that resource, or by threats of action. In the second case, resource degradation may provoke a reaction by the holder of that resource or may lead to political or economic exploitation. These effects will have consequences for international relationships, behaviour, and policy[5,6,7,8,9,10]. Figure 1 shows the path by which global climatic changes, or other large-scale environmental problems can affect international politics and, ultimately, international security.

Future Climate: What Do We Know and How Do We Know It?

Growing atmospheric concentrations of carbon dioxide and other trace gases - produced primarily from the combustion of fossil fuels and from other industrial activities - are beginning to alter the heat balance and the dynamics of the atmosphere. These gases retain heat in the atmosphere that would otherwise be re-radiated into space, thus raising the temperature of the Earth. The best current studies estimate that global average temperatures will increase 3° with an equivalent doubling of atmospheric carbon dioxide concentration. (An "equivalent" doubling of carbon dioxide refers to the amount of carbon dioxide and other trace gases (methane, chlorofluorocarbons, nitrous oxide, and so on) that has the same radiactive effect as doubling carbon dioxide alone. An "equivalent doubling" of carbon dioxide will thus occur before the actual carbon dioxide concentration in the atmosphere doubles, because of the role of the other trace gases.) Figure 2 shows the likely future trend of global average temperature together with the temperature record for the past 125 years. The rise in temperature will be accompanied by many other climatic and geophysical effects, including changes in precipitation patterns, changes in the intensity and duration of storms, and a rise in sea level due to thermal expansion of the ocean and a melting of land ice. Recently, new concerns have been raised about the possibility of surprises - sudden abrupt shifts - in the way the oceans and atmosphere may respond to climate perturbation[12].

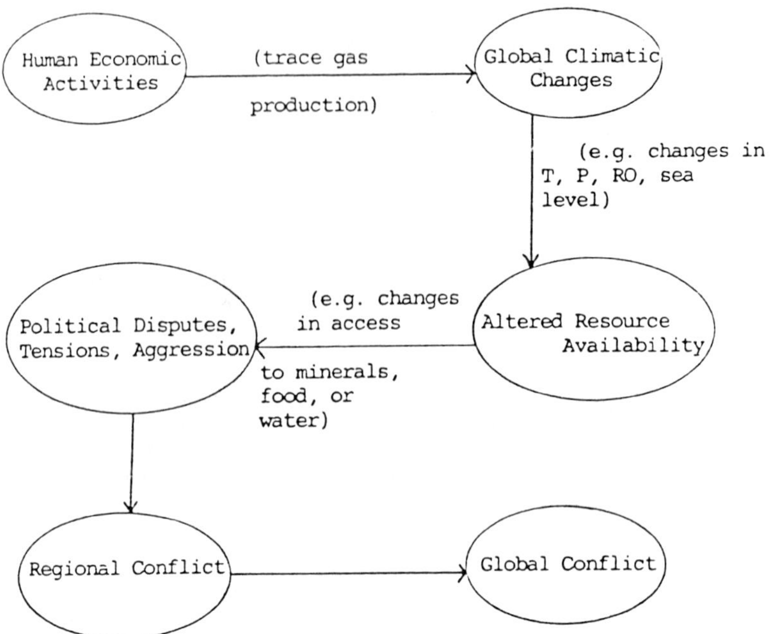

Figure 1. Environmental routes to conflict. This diagram shows the relationship between global climatic changes, impacts on resources, and international frictions and tensions. Climatic changes can act as the trigger to or the roots of international disputes and can lead to conflict when other tensions exist between states.

Because of the limitations of climate modeling, uncertainties remain about the nature, timing, magnitude, and regional details of climatic changes. Although we could wait for more research and more detailed information on the environmental and economic impacts of climatic changes before taking actions, this approach has two serious flaws. First, the complexity of climate modeling means that the necessary research will be slow and difficult. Yet, unless actions to prevent the most significant climatic changes begin soon, the earth will be irreversibly committed to substantial warming. Second, any international agreement to prevent major climatic changes may be complicated by a desire of certain actors (countries, subnational groups, corporations, alliances) to capitalize on perceived regional advantages in the midst of larger-scale negative effects. Thus, those actors who perceive themselves to be beneficiaries of a warmer earth will have an incentive to be uncooperative, even if that perception turns out later to be wrong.

The idea that some actors will benefit from climatic changes while others lose must be scrutinized carefully, because such ideas often help to drive policy actions and decisions. Is there clear evidence to suggest that regions, nations or sub national groups could benefit from long- or short-term climatic change? Climate impact research over the last few years has identified both regional advantages and disadvantages of greenhouse warming. These impacts may only be the more easily identified changes; they may not include all the most important consequences. Furthermore, any assessment of climatic impacts is enormously complicated by the different measures used to evaluate impacts and by the non-quantifiable nature of important ecological

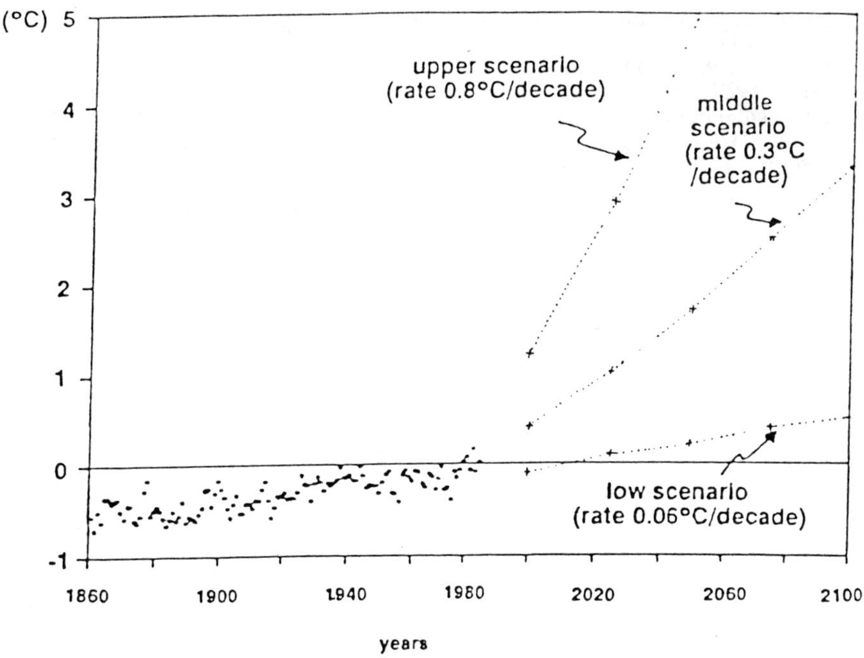

Figure 2. Three scenarios of global average temperature that might develop in response to growing atmospheric concentrations of greenhouse gases, plotted as differences from the average 1985 temperature. The middle scenario reflects present emission trends (except for limits on chlorofluorocarbons as agreed to in the Montreal Protocol). There is a 50% chance that the actual path of climatic change could lie above the middle curve. In the judgement of WCP [11] there is a 90% chance that the actual future temperature rise will lie between the upper and lower curves.

and societal disruptions. Given the inadequate state-of-the-art of climate impact research, perceptions of winners and losers are as likely to be wrong as right.

Additional complications will arise from climate impacts that may initially be favourable, but than become unfavourable as climatic conditions become more severe. For example, recent research[13] show that transient changes in climate in northern regions of the Soviet Union may benefit agriculture in the short-term, but lead to longerterm productivity losses. In that study, yields of winter rye initially increase because of higher temperatures and unchanged summer precipitation, but then decrease after 2010 as global temperatures continue to rise, summer precipitation increases, and soil fertility declines. Yield reductions may be further exacerbated by increases in surface water pollution from nitrate leaching. A massive effort to shift agriculture production north could therefore end in failure and the mis-direction of limited resources that could more effectively have been put into other agricultural projects.

Developed and Developing Countries: Roles and Responsibilities

Three fundamental characteristics of the climate issue will affect both perceptions of climate impacts and international reactions. First, the industrialized countries are primarily responsible for the production of greenhouse gases. Second, the consequences of climatic changes will be far more widely distributed among both rich and poor countries. And finally, the ability of countries or regions to react and adapt to climatic changes varies widely. These characteristics may play as much of a role in determining the extent to which tensions and conflicts develop over climatic changes as the details of the impacts themselves.

The industrialized world is primarily responsible for the production of greenhouse gases. Table 1 shows the high percentage of carbon dioxide and chlorofluorcarbons (CFC) produced by the major industrialized countries in the mid-1980s. As the Table shows, the northern hemisphere industrialized countries are currently responsible for about 70 per cent of the global production of carbon dioxide. The United States alone is responsible for nearly 25 percent of all CO_2 production and 20 to 30 per cent of CFC production. When these data are combined with population data, the discrepancy between the industrialized and developing worlds becomes even more apparent. Figure 3 plots the per capita CO_2 production as a function of population for many of the world's countries and regions.

Table 1

Per cent of World Production of Carbon Dioxide and Chlorofluorocarbons

	CO_2*	CFC-11**	CFC-12
United States	24	23	30
Western Europe & Canada	16	49	33
Japan, Australia, New Zealand	6	20	15
Soviet Union and other CPEs	26	6	19
All others	28	2	3

Notes

* These figures include carbon dioxide from fossil fuel combustion and industrial fuel use, but do not include the contribution from biomass burning[14,15].

** Chlorofluorocarbon data are from Mintzer[16].

Unlike the distribution of responsibility for greenhouse gases, the distribution of impacts from greenhouse warming will be global in extent and delayed in time. Among the victims of climatic change will be those nations (and generations) least responsible for the production of greenhouse gases, least able to either adapt to or mitigate the changes, and with little international economic or political clout. This equity problem, alone, will worsen international frictions and disputes.

Five areas stand out as particularly important for national and international relationships, behaviour, and policy: agricultural productivity and trade, the availability and quality of freshwater resources, access to freshwater resources, access to energy resources, sea-level rise, and ocean/ice conditions in the high latitudes.

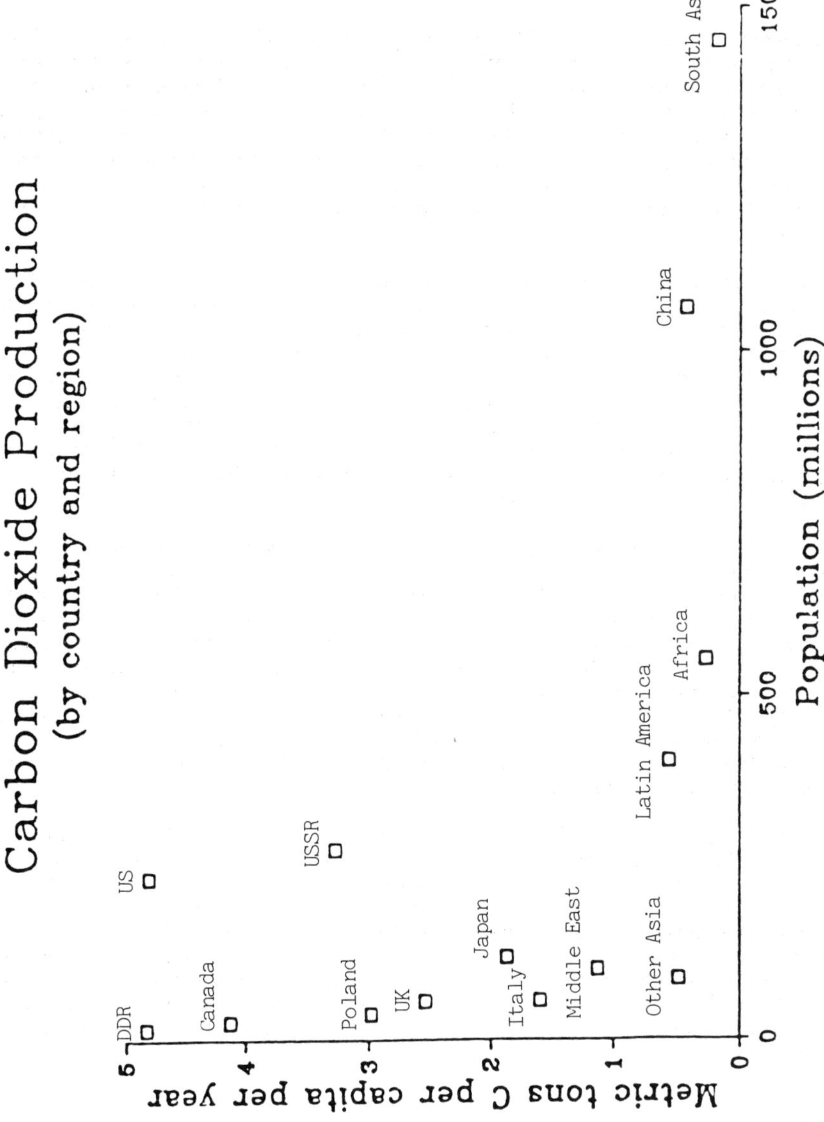

Figure 3. Per capita production of carbon dioxide (in metric tons of carbon per capita) by country or region for 1983 versus population (in millions). Note the high per-capita annual carbon dioxide production for the industrialized countries with low populations, while China, Latin America, Africa, and other developing countries and regions produce little carbon dioxide per capita.

Agriculture

Threats to the basic food supplies of a country are cause for frictions and tensions between nations[4,17]. Possible mechanisms for such threats include trade embargoes or other forms of political manipulation of access to food, environmental degradation such as loss of soil fertility, or competition among conflicting land uses. Because scarcity is a fundamental condition for a good to become a political tool, the disparity in food needs and food resources between the developing and the developed world has long hinted at the possibility of future conflict over access to food resources.

Food availability depends on a complex array of factors, including patterns of production, purchasing ability, and the operation of food-distribution systems. Many developing countries are acutely vulnerable to natural climatic variability that may cripple their own food production or substantially reduce the supply and raise the price of foodstuffs on the world market. Rainfed agriculture, in particular, is sensitive to the vagaries of weather, especially the timing and length of the rainy seasons and the total amount of rain received. Under conditions of higher temperatures, changing rainfall patterns, and growing population, the situation for some developing countries may grow more precarious.

The situation for large grain producers is equally uncertain. As temperatures increase, irrigated agricultural production could expand into northern regions of the United States, the Soviet Union, China, and Canada only if soil conditions, water availability, and other factors (such as infrastructure) permit. But output in regions that are currently productive, such as the Central Plains of the United States, the Ukraine, and Kazakhstan, could be reduced by higher temperatures and changes in water availability. Similarly, a near-term improvement in yield, due to higher temperatures or greater water availability could be followed by a longer-term deterioration of yield, as temperatures continue to rise, or as beneficial rainfall patterns disappear.

Analysis of the net effect (both regionally and globally) of climatic changes on food production is further complicated by the difficulties of estimating the effect of changes in yields on world agricultural markets. Short-term reductions in yields alone are not necessarily bad for overall long-term productivity and food availability. Confounding factors include the size of stocks, subsequent investments in other regions, planting patterns, international prices, and the character and mechanisms of trading agreements[18,19]. Thus, changes in the comparative advantage of the different actors can play a greater role than absolute changes in agricultural productivity.

Water resources

Historically, international political frictions and tensions have arisen over the control of, access to, or the quality of freshwater resources[4,10,20,21]. Even in the absence of climatic changes, pressures on existing water resources are growing due to increases in population, industrial water demand, and development. Where water resources are shared, as in international river basins or bodies of water bordering more than one country, the possiblity of friction and conflicting demands exists.

Freshwater resources are widely shared. Nearly 50 countries on four continents have more than three-quarters of their total land area falling within international river basins, while globally 47 per cent of all land

Table 2

Rivers with Five or More Nations Forming Part of the Basin

	≠	Area (km²)
Danube	12	817 000
Niger	10	2 200 000
Nile	9	3 030 700
Zaire	9	3 720 000
Rhine	8	168 757
Zambezi	8	1 419 960
Amazon	7	5 870 000
Mekong	6	786 000
Lake Chad	6	1 910 000
Volta	6	379 000
Ganges-Brahmaputra	5	1 600 400
Elbe	5	144 500
La Plata	5	3 200 000

area falls within international river basins[22]. Over 200 river basins are multinational, including 57 in Africa and 48 in Europe. Table 2 lists thirteen different major rivers with five or more nations forming part of the watershed[22].

Regions with a history of international tensions or competition over water resources include the Jordan and Euphrates rivers in the Middle East, the Nile, Zambezi, and Niger rivers in Africa, the Ganges in Asia, and the Colorado and Rio Grande rivers in North America. As water demands increase, the probability of conflict over remaining water resources will also increase. In these cases, institutional mechanisms for allocating shortages must be designed and agreed upon before climatic changes begin to alter the hydrologic regime. International water law, though immature in its development and application, must play a role[10]. By codifying these allocation schemes, mechanisms can be put into place to resolve water conflicts before they become acute.

Northern energy resources

Major energy resources underlie Arctic continental and offshore regions and these resources play a vital role in national economies and world economic markets. Any change in climate that affects the ability to extract these resources will be a factor in the response of international actors to initiatives to control climatic change. Access to northern oil and gas is already constrained in some regions by climatic conditions and it is dependent on expensive and vulnerable methods and materials.

The oil and gas potential of the northern Arctic regions is large. Despite limited exploration, over 20 per cent of the "proved" reserves of the United States are in Alaska, and nearly 40 per cent of Soviet proved reserves are in northern and western Siberia. Table 3 shows the Arctic proven reserves of the principal oil-producing countries of the region[23,24,25].

Given the importance of northern energy resources, climatic constraints are unlikely to prevent future development. Even today, despite the

Table 3

Arctic Oil Resources

Country	Arctic Proved Reserves (10^9 billion barrel)	National Proved Reserves	Arctic Proved Reserves as a Percentage of National Reserves
United States	7	33	21
Soviet Union	22	59	37
Canada	1 - 2	8	12 - 24
Norway	(*)	11	(*)

* Accurate Norwegian oil reserves data are unavailable, but estimates range between 100 and 300 million tonnes (or 0.7 to 2 billion barrels).

difficulties in exploiting northern minerals, the Western Siberian oil fields produce about half of all Soviet oil[26]. Future climatic changes may, however, significantly increase or decrease the difficulty, and hence the expense, of the production.

Although problems of access to northern energy resources do not directly affect developing countries, there are a number of indirect effects worthy of more detailed analysis. How, for example, will world energy markets be affected by easier (or more difficult) access to northern oil and gas? Will the world market price of fossil fuels be affected? Certainly, if exploiting these resources becomes easier, their cost will decrease and their use may increase. Perversely, this could further exacerbate the greenhouse effect by encouraging more combustion of carbon-based fuels.

Sea-level rise

One of the most pressing direct effects of greenhouse warming will be rising sea-level due to both thermal expansion of the oceans and melting of glaciers and land ice. While the rise in sea-level will most likely be incremental, the damages caused by the rise will occur during specific events such as storm surges, hurricanes, and typhoons. Although the impacts of sea-level rise will vary depending on the geophysical characteristics of the region, some observations can be made about the differences between developing and developed countries.

In developing countries, coastal populations and coastal agricultural developments are particularly vulnerable to storms. Other than coastal cities, sea-level rise will be most severely felt in deltaic areas, especially in countries where rivers are large and deltas extensive and highly populated. The highly-populated coastal plains of India and Bangladesh, for example, are already extremely vulnerable to storms. Since 1960, India and Bangladesh have been struck by at least eight tropical cyclones that each killed more than 10 000 people[27] (see Table 4). In late 1970, storm surges killed approximately 300 000 people in Bangladesh and reached over 150 kilometres inland over the lowlands. Recent estimates show that a climatically-induced one-metre sea-level rise would cover land presently occupied by nearly 10 per cent of Bangladesh's population[28]. A far greater fraction of the population would be put at risk by the increased consequences of storms.

Table 4

Major Storm Disasters in Coastal/Delta Regions

Year	Location	Estimated Death Toll
1963	Bangladesh	22 000
1965	Bangladesh	17 000
1965	Bangladesh	30 000
1965	Bangladesh	10 000
1970	Bangladesh	250 000 - 500 000
1971	India	10 000 - 25 000
1977	India	10 000
1985	Bangladesh	10 000

The delta in Bangladesh is also vulnerable to the activities of upstream states: India, Nepal, China, and Bhutan. Disputes already exist between India and Bangladesh over the construction of dams on the Ganges, and over the use of major upstream tributaries. Upstream activities can affect both water availability in Bangladesh and the formation and ecology of the delta, which offers some protection from storm surges and sea-level rise. One solution, population migration, offers a way of reducing the human vulnerability in some areas, but at the high societal cost of relocation and absorption of those displaced.

Of greater concern in the developed world is the protection of coastal cities and industrial sites by building structures to protect against sea-level rise. In the Netherlands - a small but vulnerable country - over half the country would be uninhabitable if not protected against the sea[29] and 8 million people would be displaced. In the United States, a mixed strategy of abandonment and protection is being considered, with attention being given almost exclusively to threats to cities.

Arctic and antarctic ocean and ice conditions

Arctic and Antarctic conditions are beginning to play an important role in both strategic and commercial fields. In the Antarctic, questions about access to mineral resources are beginning to re-emerge as we near the 30th anniversary of the Antarctic Treaty of 1959, and as capabilities for minerals extraction under severe conditions improve. Despite several years of discussion, no treaty that would regulate or prohibit the exploitation of Antarctic mineral resources has yet been worked out. As climatic conditions begin to alter extraction conditions in the higher latitudes, disputes could become more prevalent, especially since both developed and developing countries maintain unresolved sovereignty claims in the Antarctic.

In the Arctic, commercial sea routes in the Northeast Passage by the Soviet Union are kept open at considerable expense by Soviet icebreakers. As the world warms, the Arctic could become ice-free for a large part of the year. The implications of this for commercial trade are unclear, but one important consequence could be the reliable opening of the Northeast Passage over the Soviet Union and the possible opening of a satisfactory Northwest Passage over Canada.

There are also implications for the strategic balance between northern hemisphere superpowers. The Arctic is considered to be an area of strategic importance to the United States and to the Soviet Union[30], yet activities there are severely hindered by environmental conditions. This ice-covered region provides cover for the actions of nuclear-powered submarines of the United States and the Soviet Union, over the occasional objections of the other bordering nations - particularly Canada. While it is difficult to predict the importance of strategic submarines (or even the relations between the United States and the Soviet Union) in 25 to 40 years, the possibility of much less severe ice conditions in the Arctic could have important security ramifications.

The Ability to Respond or Adapt to Climatic Impacts

There is a fundamental discrepancy between the ability of developed and developing countries to adapt or respond to the impact of climatic change. In some cases, climate impacts may be more severe in countries that are the least able to adapt. For these countries, prompt preventative measures are far preferable. Yet, the burden for prevention rests with those countries that bear the greatest responsibility for the production of greenhouse gases. Where the severity of the impacts are uncertain, developed countries are likely to lean toward adaption. One risk of this strategy, of course, is that unexpectedly severe or rapid impacts may overwhelm the ability to adapt, or that there are thresholds or nonlinearities in societal capabilities for adaption[18].

In the absence of prompt and severe climatic impacts associated with greenhouse gases, industrialized nations are likely to prefer the "wait and see" strategy of adaptation, which produces the fewest near-term costs to society. The corollary to this, however, is that reliance solely on adaptation may produce the greatest long-run costs from climatic impacts. Developing countries with limited resources for responding to major climatic changes, such as sea-level rise, loss of agricultural productivity, and so on, have the most to gain by forcing greenhouse gas producers to cut back on carbon dioxide production.

A further complication may be a desire of certain actors to capitalize on perceived regional advantages of climatic changes. Those who believe, rightly or wrongly, that they will benefit from a warmer earth will have no direct incentive to cooperate in any international agreement to prevent climatic change. If these actors are among those most responsible for greenhouse gas production - and hence most able to affect the outcome - others may be forced into an adaptation strategy.

Conclusions

Of all the pressing large-scale environmental problems facing society, global climatic changes appear to have the greatest potential for provoking disputes, worsening tensions, and altering international relations. The impacts associated with global warming, changes in precipitation and storm patterns, and rising sea level will be felt throughout the world.

A disproportionate share of the responsibiltiy for climatic changes lies with the industrialized countries. The societal impacts of these climatic changes, however, will be far more equally distributed, posing important and still unresolved questions about equity and fairness. With few exceptions, developing countries are least able to mobilize the resources for adapting to severe impacts, while developed countries are more likely to prefer to adapt to climatic changes than to alter drastically their energy and industrial infrastructures. This fundamental dichotomy is likely to lead to international frictions and disputes between developed and developing countries, unless mechanisms can be worked out to share more equitably the costs of climatic impacts and the costs of preventing them.

Both technical and political solutions are available, although none appear capable of preventing at least some climatic changes. Among the technical options are (a) prevention - or at least slowing - of climatic change through increased energy efficiency, greater use of non-fossil-fuel energy sources, reforestation, controls on other greenhouse gases, and the sequestering of carbon dioxide away from the atmosphere, such as deep in the ocean; (b) mitigation of the most severe impacts through larger grain storage, more flexible water resource systems, increased water conservation, and so on; and (c) adaptation to climatic impacts such as through population migration, land abandonment, shifts in agricultural practices, and selective protection of coastal property. Indeed, the actual responses to climatic changes are likely to include all of the above.

Various political actions are also available depending on the perception of the severity of climatic impacts and on the economic and technological means available for responding. Unilateral actions can be taken by those countries that have a major reponsibility for greenhouse gas production, or that face particularly severe climatic impacts. Although unilateral actions are likely to be only marginally effective at preventing climatic changes, they set good examples and some of them can be justified for other good reasons. For example, increasing the robustness of the exisitng agricultural and water-resource systems would ease many existing food and water problems, while helping to prepare for similar problems caused by future changes in climate. Increased energy efficiency reduces the economic, political, and environmental costs of energy use while slowing the emissions of carbon dioxide. Restrictions on ozone-destroying chemicals protect human health and slow the rate of climate change.

Even more effective would be multi-lateral actions on the part of the industrialized world. Coordinated actions to control trace-gas emissions should be the responsibility of these countries and would help address the questions about equity raised above. Finally, comprehensive international agreements, such as those developed for protecting the ozone layer, or the Law of the Sea, would be most effective at slowing the rate of climatic change, but, of course, the most difficult to negotiate and implement.

There are no simple solutions to the problem of large-scale climatic change - in fact, the Earth is already committed to some warming because of gases already in the atmosphere. Nevertheless, even slowing the rate of change by slowing the rate of production of the major trace gases can provide considerable "breathing room" to allow us to both improve our understanding of climatic impacts and to reflect on equitable international responses.

References

1. World Commission on Environment and Development (WCED), Our Common Future, Oxford, England: Oxford University Press, 1987, 400 pp.

2. R.H. Ullman, "Redefining Security", International Security 8, No. 1, 1983, pp. 129-53.

3. A.H. Westing (editor), Global Resources and International Conflict: Environmental Factors in Strategic Policy and Action, Stockholm International Peace Research Institute, United Nations Environment Programme, Oxford: Oxford University Press, 1986, 280 pp.

4. P.H. Gleick, "The Implications of Global Climatic Changes for International Security", Background Paper No. 14 for the Conference on Developing Policies for Responding to Future Climatic Change, Villach, Austria, September 28 - October 2, 1987.

5. M.I. Budyko, Climatic Changes, (Translation by the American Geophysical Union, Baltimore, Maryland: Waverley Press, Inc., 1977.

6. W.O. Roberts and H. Lansford, The Climate Mandate, San Francisco, California: W.H. Freeman and Company, 1987, 197 pp.

7. T. Gustafson, Reform in Soviet Politics: Lessons of Recent Policies on Land and Water. Cambridge : Cambridge University Press, 1981, 218 pp.

8. T.W. Wilson, Jr., "Global Climate, World Politics and National Security", in World Climate Change: The Role of International Law and Institutions, V.P. Nanda (editor), Boulder, Colorado: Westview Press, 1983, pp. 71-8.

9. C. Tickell, Climatic Change and World Affairs, revised edition, Center for International Affairs, Harvard University, Lanham, Maryland: University Press of America, 1986, 76 pp.

10. P.H. Gleick, "The Effects of Future Climatic Changes on International Water Resources: The Colorado River, The United States, and Mexico", in press, Policy Sciences, 1988.

11. World Climate Programme (WCP), "Developing Policies for Responding to Climatic Change", A summary of discussions and recommendations for the Workshops held in Villach and Bellagio, September-November 1987, World Meteorological Organization, Geneva, Switzerland, 1988.

12. W.S. Broecker, "Unpleasant Surprises in the Greenhouse?", Nature 328, No. 6126, 1987, pp. 123-26.

13. S. Pitovranov, V. Iakimets, V. Kiselev and O. Sirotenko, "Effects of Climatic Variations on Agriculture in the SubArctic Zone of the USSR", 97 pp, in The Impact of Climatic Variations on Agriculture, Volume 1. Assessments in Cool Temperate and Cool Regions, M.L. Parry, T.R. Carter and N.T. Konijn (editors), Reidel, Dordrecht, The Netherlands, 1987.

14. R.M. Rotty, "Estimates of Seasonal Variation in Fossil Fuel CO_2 Emissions", Tellus 39B, No. 1-2, 1987a, pp. 184-202.

15. R.M. Rotty, "A Look at 1983 CO_2 Emissions from Fossil Fuel (With Preliminary Data for 1984)", Tellus 39B, No. 1-2, 1987b, pp. 203-8.

16. I.M. Mintzer, A Matter of Degrees: The Potential for Controlling the Greenhouse Effect, World Resource Institute, Washington, D.C., 60 pp.

17. P. Wallensteen, "Food Crops as a Factor in Strategic Policy and Action", in Global Resources and International Conflict: Environmental Factors in Strategic Policy and Action, A.H. Westing (editor), Stockholm International Peace Research Institute, United Nations Environment Programme, Oxford: Oxford University Press, 1986, pp. 143-58.

18. W.C. Clark, "On the Practical Implications of the Carbon Dioxide Question", International Institute for Applied Systems Analysis, Laxenburg, Austria, IIASA-WP-85-43, 1985, 84 pp.

19. R.S. Chen and M.L. Parry (editors), "Policy-Oriented Impact Assessment of Climatic Variations", International Institute for Applied Systems Analysis, Laxenburg, Austria, IIASA-RR-87-7, 1987, 54 pp.

20. T. Naff and R.C. Matson (editors), Water in the Middle East: Conflict or Cooperation, Boulder, Colorado: Westview Press, 1984, 236 pp.

21. M. Falkenmark, "Fresh Waters as a Factor in Strategic Policy and Action", in Global Resources and International Conflict: Environmental Factors in Strategic Policy and Action, A.H. Westing (editor), New York: Oxford University Press, 1986, pp. 85-113.

22. United Nations, Register of International Rivers, Oxford: Pergamon Press, 58 pp.

23. J.N. Garrett, "Conventional Hydrocarbons in the United States Arctic: An Industry Appraisal", in United States Arctic Interests: The 1980s and 1990s, W.E. Westernmayer and K.M. Shusterich (editors), New York: Springer-Verlag, Inc., 1984, pp. 39-58.

24. H.O. Bergesen, A. Moe and W. Ostreng, Soviet Oil and Security Interests in the Barents Sea, London: Frances Pinter, 1987, pp. 144.

25. British Petroleum (BP), BP Statistical Review of World Energy, British Petroleum Company, Britannic House, Moor Lane, London, 1987, p. 36.

26. V.L. Mote, "Environmental Constraints to the Economic Development of Siberia", in Soviet Natural Resources in the World Economy, R.G. Jensen, T. Shabad and A.W. Wright (editors), Chicago: The University of Chicago Press, 1983, pp. 15-71.

27. J.E. Bardach "Global Warming and the Coastal Zone: Some Effects on Sites and Activities". For the Conference on Developing Policies for Responding to Future Climatic Changes, September 28 to October 2. Villach, Austria, 1987, p. 80.

28. J. Broadus, J. Milliman, S. Edwards, D. Aubrey and F. Gable, "Rising Sea Level and Damming of Rivers: Possible Effects in Egypt and Bangladesh", in Effects of Changes in Stratospheric Ozone and Global Climate: Volume 4, Sea Level Rise. J. Titus (editor), U.S. Environmental Protection Agency, Washington D.C. and United Nations Environment Programme, 1986, pp. 165-89.

29. T. Goemans, "The Sea Also Rises: The Ongoing Dialogue of the Dutch with the Sea", in Effects of Changes in Stratospheric Ozone and Global Climate, Volume 4: Sea Level Rise, J. Titus (editor), Washington, D.C: United States Environmental Protection Agency and United Nations Environment Programme, 1986, pp. 47-56.

30. G.L. Johnson, D. Bradley, and R.S. Winokur, "United States Security Interests in the Arctic", in United States Arctic Interests: The 1980s and 1990s, W.E. Westermeyer and K.M. Shusterich (editors), New York: Springer-Verlag, 1984, pp. 268-94.

International Cooperation for Survival

Maurice Strong

Introduction

The United Nations Conference on the Human Environment held in Stockholm, Sweden in 1972 reminded us that we live on "Only One Earth" and that we had to begin to direct more attention to caring for and maintaining the evironment and natural resource base on which the survival and well-being of the entire human family depends.

The Report of the World Commission on Environment and Development presented to the last session of the General Assembly of the United Nations by the Chairman of that Commission, Mrs Brundtland, Prime Minister of Norway, makes it clear that we must now move from one earth to one world if we are to secure our common future on this planet. It pointed out that the future of our species can be an ominous one, in which our very survival is at risk if we continue on our present course. But it also makes clear that it is entirely possible to achieve a more secure, more abundant and more hopeful future for the whole human family in the next century, if we can learn to manage more rationally and effectively our relationships with each other and our impacts on environmental and natural resource systems of our planet.

The next two decades will be the most challenging, dangerous and decisive period in the entire history of the human community. What we do, or fail to do, during this period will largely determine the future of human life on our planet.

Problems of Overpopulation

The awesome statistics which illustrate the exponential growth of the world's population are all too familiar - from 1.6 billion at the beginning of this century to 5 billion now, with the prospect of another billion being added by the end of the century. But the conclusions we draw from this are often far too simplistic. Those of us in the industrialized world in which growth has already leveled off, find it all too easy to point to the continuing explosive population growth of the developing world as the principal source of the world's problems. Yet, such major risks to the future of the whole human family as ozone depletion and man-induced climate change, stem largely from the activities of the industrialized countries. Indeed, the vast increases in industrial production and individual consumption in the rich countries has contributed far more to global environmental threats than the rapid growth of population in developing countries. But this in no way ameliorates the urgent need for developing countries to bring their high rates of population growth under control. For it threatens to frustrate their ability to improve the lot of their people through economic and social development. And it will make it difficult if not impossible for them to participate fully in the international efforts required to ensure the environmental integrity and security on which the whole human family depends.

The intensity of human activity has accelerated even more rapidly than the growth in human numbers. In this century, while population tripled, gross world product increased more than 21 times. The use of fossil fuels grew nearly 30-fold and industrial production by more than 50 times.

A revolution in agriculture has enabled food production to outstrip population growth and produce surpluses of agricultural products at a time when many had predicted that the world would be facing food shortages. Paradoxically, hunger and malnutrition are endemic today, not because of food shortages but because of the dire poverty that denies the poor access to the food they need.

The industrial and agricultural revolutions have produced unprecedented rates of economic growth and prosperity, the benefits of which have primarily accrued to the people of the industrialized countries and a privileged minority in the rest of the world. Yet, the destructive byproducts of these activities, the degradation of the environment, destruction of renewable resources and risks to human health must be borne by the entire human populttion, and most of all by the poor.

These changes have been taking place at levels of population and of human activity which are a good deal less than they will be in the next century. Despite the encouraging reductions in population growth which are occurring in many areas today, the world's population will probably double by the time it stabilizes sometime towards the middle of the next century. And the intensity of human impact on the environment and resource base will grow at an even faster rate.

Industrialized countries, in which most economic activity is still concentrated, account for only about one-quarter of the world population today, and by the beginning of the next century this proportion will be about one in five. The four out of five living in the developing world will not be satisfied to remain at the present low levels of economic development which confine the majority of them to a state of poverty that is no longer either acceptable or inevitable. Economic growth is an imperative for developing countries, and this will greatly increase their claims on world resources as well as their impacts on the earth's environmental and life support system.

Developing countries have made impressive progress in increasing their share of the world's industrial capacity. But they still have a very long way to go. It is estimated that world industrial output would have to be increased by a factor of 2.6 to raise the **per capita** consumption of manufactured goods by developing countries to present industrialized country levels. Energy production will have to be increased by a factor of 8 to bring developing countries up to current levels of energy consumption in industrialized countries.

Already industrial and urban growth in developing countries is producing the same kind of environmental problems which were once thought to be a disease of the rich. The most polluted environments in the world today are in the developing world, where cities like Manila, Cairo, Mexico City and Santiago are experiencing water and air pollution problems amongst the world's worst.

Stabilization of world population is a critical imperative for the next century. This primarily depends on reduction of population growth rates in developing countries. But population policies clearly will only work when the motivations of individuals correspond with the interests of society as a whole. And there is abundant evidence that economic and social development provides the best motivation for effective population control.

As the Report of the World Commission on Environment and Development makes clear, a transition to sustainable development is the only viable pathway to the new era of economic growth which the world requires. This means achieving a positive synthesis between economics and environment in the establishment of our economic policies, and in managing their implementation in every sector of our economic life. It will mean, too, a massive shift of economic growth towards the developing countries. While this is bound to be disruptive and difficult for the traditional industrialized countries, it will ultimately provide them with greater economic opportunity as well as security. Translating the immense unmet needs of the people of the developing world into markets represents the best prospect of achieving a new era of sustained economic growth in the world economy as a whole.

Industrialized countries must lead the way in effecting the transititon to sustainable development. Fortunately, some of the processes already underway are moving us in this direction. Under the impetus of higher energy prices and potential resource shortages, the raw materials content of industrial production has been reduced dramatically in industrialized countries during the past few years. Japan, for example, used only 60 per cent as much raw materials for every unit of industrial production in 1984 as it used in 1973. The principal source of added value today is the application of human knowledge and intelligence to products and services. These must be directed more and more to meeting the non-material needs and aspirations of people in the fields of culture, education, spiritual development and recreation.

A related phenomenon is the internationalization of the processes of industrial production. More and more products today are manufactured using components from a number of different countries which can make them most competitively. The internationalization of our financial, monetary and trading systems has already advanced to the point where they are global in both the nature and scale of their operations. The recent turbulence in the stock market has demonstrated how quickly changes of mood and investment behaviour in one country can be transmitted through the whole world's financial systems.

The processes of internationalization in the economic sphere have important implications, too, for the prospects for peace and security. Global military spending now equals some $1000 for every one of the world's billion poorest people. Military expenditures are impoverishing many of the poor countries of the world; but they are also weakening the economies of the superpowers and provide a powerful incentive for the reduction in armaments.

The operation of economic interdependence has reduced immensely the capacity of nations to wage a modern war. France and Germany could not now sustain a war against each other; neither could Japan sustain a major war. Surely the best hope for peace in the world is to make war unfeasible. This, in my view, is one of the most hopeful and promising products of interdependence.

Interdependence is not an unmitigated blessing for all. It often multiplies the risks and the vulnerabilities faced by the weak and the poor. This was dramatically manifest in sub-Saharan Africa where high energy prices followed by world recession, high interest costs and sharp declines in commodity prices compounded the effects of drought and ecological breakdown to produce the massive human tragedy of famine.

Today the stresses and strains and problems created by interdependence are more in evidence than its positive and hopeful features. There is a resurgence of parochialism and its economic counterpart, protectionism, accompanied by assertion of the kinds of narrow, short-sighted, national self interests which have proven so destructive in the past.

The processes by which the future of our planet is being shaped are essentially systemic in nature. Each one of the elements of these changes affects and is affected by the other. Industrial practices and consumer behaviour in industrialized countries give rise to global environmental risks. Agricultural subsidies in the USA and the Common Market undermine food production in Africa and even contribute to the march of the deserts. Population growth leads to destruction of renewable resources which in turn leads to poverty and famine. Competition between the superpowers nourishes conflicts in developing areas which perpetuate their poverty and add to their suffering. Escalating military expenditures everywhere drain away the resources which could be used to relieve poverty, reduce environmental destruction and facilitate sustainable development.

Although virtually all of the human activities which give rise to these conditions take place within nation states, and most are rooted in local and individual actions, all are part of a complex system of interacting relationships which transcend the boundaries of nations. The real world cause-and-effect system, in which human activities interact to produce the results that determine our future, is inherently global in its dimensions. What happens in any part of the system can affect the whole.

New Institutional Mechanisms

The awesome advances in science and technology which have made possible our unprecedented economic progress have not been accompanied by corresponding progress in development of the political will and institutional mechanisms required for effective control and management of the vast powers which these advances have placed in our hands. While science and technology have propelled us into a new and exciting era in which future visions are rapidly becoming current realities, our political attitudes and institutions are still rooted stubbornly in the past.

The institutions and processes through which we manage our affairs at every level of society must be structured and geared so as to interact effectively with the real world cause-and-effect processes they are designed to manage. They must therefore be systemic in nature. The often tightly defined, almost always jealously defended boundaries between various levels of jurisdiction and sectoral activity within each country and each institution, will have to give way to more integrated, systemic management structures.

The private sector has already moved far beyond government in creating and managing the rapidly evolving global world economic and financial systems. The processes of industrial production, the financial system and the trading system have become internationalized to an unprecedented degree. Interdependence is becoming a reality and multinational corporations are its most effective agents. Governments are far behind. The mechanisms of governance at the international level are the most primitive and least developed of all.

The United Nations, the principal global organization, has special handicaps. Its powers are always subject to the veto of any one of the five permanent Security Council members which can, and often do, override the will of the majority of countries. The wealthy countries which provide most of its funding wield an additional **de facto** financial veto. This was dramatically manifest recently when the United States chose to withhold some of the funding of the United Nations to which it is committed by its treaty obligations. And the fact that the specialized agencies of the United Nations operate, for all practical purposes, independently of the central body of the UN is a major constraint on its capacity to provide a framework for effective systematic management at the global level.

An effective global level of governance need not and should not be a world government - a supreme authority for the control and direction of all forms of human activity. It should deal first and foremost with those issues which bear to an important degree on human survival and well being, and which inherently require cooperation beyond the boundaries and juristictions of individual nation states. This includes issues such as care and use of the global commons, anticipating and dealing with the causes and effects of major changes in climate, control of trans-boundary pollution, management of the world economic, financial, monetary and trading systems, and the maintenance of world peace and security.

In my view, the organizing principle which is best applied to any system of governance or management is that responsibility for every activity should be vested in the level closest to the people affected at which it can be managed effectively. By this criterion, the number of issues that would be dealt with at the global level would be relatively limited, much more limited than the currently overcrowded agenda of the United Nations. Of highest priority are those issues which affect the basic "outer limits", or boundary conditions, which are essential for the security and well being of the entire human family. Within these boundaries, an infinite variety of individual and national activities can take place, and these can be dealt with largely by other levels of governance and, of course, the private sector. Even in these cases, the context will often be global, and illuminating that context will be an essential function of the global level.

The wide variety of organizations and institutions which make up the private sector - corporations, voluntary agencies, universities, trade unions and citizens' groups - will become even more important participants in the processes of governance at every level. And many of them, notably private sector corporations, are far ahead of government in responding to the requirements and opportunities of the technological age. But private sector organizations will have to accommodate, too, to a much greater degree of coordination and harmonization of their activities with governments, international organizations and other actors. And we are only at the very early stages of developing the attitudes and the mechanisms required to accomplish this.

Cultural Dimension of the Changes

Changes of the magnitude required to manage our future successfully cannot be accomplished solely at the economic and technocratic levels. They must respond to the deepest motivations and instincts of people. The cultural dimensions of the changes taking place in our world society are of profound importance. The human community consists of a multitude of cultures of astounding variety, each with its own language and deeply rooted traditions and values. In recent times this has come to be covered with a thin but growing pervasive layer of universal and universalizing culture. The English language, rock music, Coca Cola, fast foods, Hilton hotels and international airports and conference centres are important manifestations of this. For us Westerners, it is relatively easy to live in this universal culture, as it is largely an extension of our own. But for the people of other cultures who must come to terms with it, it gives rise to apprehensions, resistance and tensions.

The world need not, and should not, become a universal monoculture. In nature we learn that there is great strength in diversity and variety. The same is surely true for the human community. The universal culture which is now emerging must not and need not smother traditional cultures and values. Rather it must provide a framework in which they can continue to flourish.

The universal culture must go well beyond the economic and materialistic manifestations which have principally characterized it to date. One of its principal elements must be an ethic of cooperation which will mean, for the Western world in particular, a shift away from the emphasis on competition and the aggressive assertion of individual self interest which has characterized our industrialized societies. The traditional virtues which we have honoured more in lip service than in reality - caring for, sharing with, and respecting the rights of others - must be seen not just as pious ideals divorced from reality, but as imperatives for our common survival and well-being.

As human civilization has evolved, people have learned to enlarge their circles of allegiance and loyalties from the family, to the tribe, to the village, to the town, to the city, to the city state, and to the nation state. We must now take the next, and final, step, at least on this planet, to the global level. This is, in my view, the most compelling imperative that confronts us as we prepare for the next century, and the most exciting challenge.

The kind of global system of management I envisage will clearly require a degree of discipline, constraint and mitigation of sovereignty which nations, institutions and individuals are always reluctant to accept. But accepting them is essential if the vastly greater number of opportunities which interdependence makes possible are to be realized.

The role of the automobile is an example. Having to drive on one side of the road, obey traffic signals, obtain licences and observe safety standards, all in a sense restrict our individual freedom. But they are necessary for us to enjoy the vast array of options and mobility which the automobile provides. So, too, the many other constraints and disciplines we will have to accept under an effective global management regime will in turn open up a range of options and benefits for individuals and societies immensely greater than the limitations they impose on our individual and national freedoms. Thus, we must continuously seek the best possible balance between individual freedoms on the one hand and, on the other, the collective disciplines and constraints required to make it possible for these freedoms to be effectively exercised and available to all.

In many ways the challenge we face is one of reconciling the world's egosystems through which human drives and aspirations are expressed with the ecosystems of our planet which prescribe the boundary conditions we must respect to ensure our common future as a species. We have now reached the point at which the scale and intensity of human activities is the decisive factor in determining the functioning of this global cause- and effect-system. We are literally in command of our own evolution. And we cannot escape accepting the responsiblities which this imposes on us for the management of our own future on this planet.

Part VIII

Alleviating Underdevelopment

Disarmament for Development

Nicole Ball

It has long been argued that the United Nations Charter defines for the UN a central role in the promotion of both disarmament and development, and it is under the auspices of the UN that much of the discussion of the relationship between disarmament and development has occurred. While disarmament and development can and, indeed, ought to proceed independently of one another, it is clear that the two concepts are linked through the increasingly large sums of money devoted to the global military sector and the continuation of poverty and maldevelopment in the Third World. The release of funds from the military sector could, if reinvested in productive civil-sector undertakings, contribute to the increased welfare of the Third World.

> "...disarmament and development have a close and multidimensional relationship. Each of them can have an impact at the national, regional and global level in such a way as to create an environment conducive to the promotion of the other.
> ...The relationship between disarmament and development in part derives from the fact that the continuing global arms race and development compete for the same finite resources at both the national and international levels. The allocation of massive resources for armaments impedes the pursuit of development to its optimal level"[1].

Some Proposals

Over the last three decades, there have been a series of proposals which envisage applying money previously spent in the security sector to development. Some proposals have been made by peace groups and peace researchers, but the majority have been made through the United Nations system[1]. In some proposals it is suggested that only the five Permanent Members of the UN Security Council reduce their security expenditure, in others, all countries; in yet others, only developing countries. The methods for acquiring the money to be applied to development vary also. Some proposals suggest taking money saved by the conclusion of arms-control agreements while others suggest taxing arms transfers.

Three of the proposals made to the United Nations are summarized in Table 1. United Nations General Assembly Resolution 3093 of December 1973 was based on a proposal by the Soviet Union. Permanent Members of the Security Council (USA, USSR, China, France, UK) were to be required to participate, and other major economic and military powers were to be invited to adhere voluntarily. All participants were to carry out a one-time reduction of security expenditure and devote 10 per cent of the savings to Third-World development. The remaining 90 per cent could be retained by the disarmer.

United Nations General Assembly Resolution 3463, of 1975, was proposed by Sweden and Mexico. All UN Member governments, particularly the five

Table 1. UN Disarmament Development Proposals

	UNGA Res.3093[a]	UNGA Res.3463[b]	French Proposals[c]
Disarmers	Five Permanent Members UN Security Council	All States, esp. Five Permanent Security Council Members	Most heavily armed and most developed
Reduction	10% of 1973 military expenditure	To be negotiated	$1 billion voluntary contributions savings from arms-control agreements
Development Bonus	1% of 1973 military expenditure (10% of 10% reduction)	Unknown	All money to be placed in a Disarmamaent Fund for Development
Recipients	Unspecified developing countries	Unspecified	Poorest and least-developed countries

a = Proposed by the Soviet Union in 1973
b = Proposed by Sweden and Mexico in 1975
c = Proposed in 1978 and 1984

Permanent Members of the Security Council and other Governments that spent a comparable amount on the military, were encouraged to participate in a one-time reduction of security expenditure, the size of which was to be based on negotiations among the participants. It was not stated what was to be done with the money saved but presumably most, if not all, would be retained by the disarmer.

The French proposals to the United Nations for the creation of a Disarmament Fund for Development was first made in 1978. It was subsequently examined by a UN Group of Governmental Experts under the aegis of the United Nations Institute of Disarmament Research (UNIDIR) - with funding from the French government - in anticipation of the International Conference on the Relationship between Disarmament and Development that was to take place in Paris in 1986. Although this conference was postponed, the proposal was taken up when that Conference was eventually convened in New York in 1987.

The original French proposal envisaged creating the Fund with a one-time contribution of $1 billion to be collected from the "most heavily armed and most developed" countries. These initial assessments for individual countries were to be based on:

- the number of nuclear-weapon vehicles possessed
- the amount of certain types of conventional weapons possessed.

Longer-term financing for the Fund would be obtained from arms-control agreements. The notion was that money would be saved when countries agreed to limit the quantity or type of weapons in their arsenals, and that some portion of these savings could be used to finance the Disarmament Fund for Development. The beneficiaries were to be the "poorest and least-armed" countries. To a large degree, the proposal that emerged from the UNIDIR exercise was similar to the original French proposal. However, at the International Conference, the issue of creating such a Fund was sidestepped. It was simply agreed that "further consideration" would be given to the adoption of measures to reduce military expenditure and that Member Governments would "assess" the potential for releasing resources through arms-control agreements and would "consider" including in disarmament negotiations "provisions to facilitate the release of such resources"[2]. In short, it was to be "business as usual" concerning reductions in military arsenals and in military expenditure, and the transfer of funds released through such reductions to increased development assistance.

Some Problems

Each of these proposed models has one or more shortcomings which apply equally to other disarmament-development proposals not discussed here. For example, where the proposals which suggest specific reductions in security expenditure - as in the 1973 Soviet proposal - run into the problem of measuring security-related outlays.

The UN Secretary-General had mandated an Expert Group to examine the issues growing out of UNGA Resolution 3093. That Group concluded that "a prerequisite for negotiating the reduction of military expenditures was agreement on the scope and content of such expenditures"[3]. Unfortunately, many governments resist giving information on the "scope and content" of their security expenditure, and the Soviet government has been foremost among them. In 1978, the UN Expert Group on the Relationship between Disarmament and Development, established at the UN Special Session on Disarmament in June of that year, recommended to the Secretary-General that UN Member Governments provide information on their security spending broken down "in terms of personnel, administration, procurement, research and development, capital investment"[4]. Senior Soviet administrators in the UN Department of Political and Security Council Affairs were able to prevent action upon this recommendation.

The UN continued, nonetheless, to develop a unified reporting system for military expenditure. This system was tested in 1979/1980 and the first regular reporting exercise occurred in 1981. Of all WTO members, only Romania has presented any information on military expenditures. The failure of the Soviet Union to participate in the UN reporting exercises provided a convenient justification for the withdrawal in 1985 of several NATO members which had provided military expenditure data for the previous five years, including the United States and the Federal Republic of Germany[5]. The issue of arriving at mutually agreed-upon estimates of military expenditure is not a trivial one. Disarmament-development proposals which are not tied to a verifiable reduction in security expenditure run the risk of not affecting security spending at all.

In part, the USSR's reluctance to participate in the security expenditure reporting exercises has stemmed from the shortcomings in its system of price allocation and from the conscious omission of major components of security expenditure from its publicly released figures. High-ranking Soviet officials have begun to state that a new cost-accounting system has been instituted which will enable military expenditure figures comparable to those produced by NATO countries to be generated within a few years. In an interview with the New York Times in October 1987, Marshal Sergei

Akhromeyev, Chief of the Soviet General Staff, admitted that many of the items included in the standard NATO definition of military expenditure (such as arms procurement and R&D) are not included in the official Soviet military expenditure figure[6]. Similar remarks by General Secretary Gorbachev have been reported in Izvestia and Pravda.

Since procurement and R&D are estimated to account for as much as 70 per cent of Soviet military expenditure, official Soviet statistics have grossly understated actual outlays for the last 25 years. This problem has not escaped Western governments and analysts, and efforts have been made to produce more accurate estimates of Soviet military expenditure. The US Central Intelligence Agency has been the ultimate source of Western estimates. It has, however, recently become public that the CIA has also omitted six to seven categories of expenditure which contribute to Soviet defence spending, such as military-related portions of the Merchant Fleet, railroad and Aeroflot, civil defence, and military assistance to countries such as Vietnam and Cuba. If outlays in these categories were added to the CIA's figures for Soviet military expenditure, they would increase military expenditure as a fraction of GNP by 5 per cent (from just over 18 per cent to around 24 per cent).

Recent research has nevertheless indicated that these military-expenditure-as-percentage-of-GNP figures may still seriously underestimate the burden of military expenditure on the Soviet economy. The CIA has estimated Soviet GNP to be approximately 55 per cent of that of the United States, and it has attributed real growth rates for the Soviet economy each year for the past ten years. However, official Soviet economists have recently indicated that Soviet GNP may be substantially lower than the CIA had estimated and that there was little or no economic growth for much of the past decade.

With proposals that call for one-time reductions in security spending, the funding which is potentially available to Third-World countries is rather small in terms of the development needs of the Third World. What is more, because this money becomes available only once, it is impossible to fund on-going projects. If, for example, UNGA Resolution 3093 had been in effect in 1983, it would have generated approximately $4 billion in development funds for Third-World countries on a one-time basis. In 1983 alone, some $3.5 billion in Official Development Assistance (ODA) was disbursed to just three countries in the Indian subcontinent (Pakistan, $669 million; India, $1725 million; Bangladesh, $1071 million.). These are admittedly among the poorest and most heavily populated countries in the Third World. Nonetheless, if $4 billion were divided among the approximately 35 least-developed countries, each would receive about $100 million as a one-time donation. While not all least-developed countries receive as much ODA as India and Bangladesh, many receive well in excess of $100 million annually[7].

Proposals that concentrate on reducing the security expenditure of the major powers leave the issue of Third-World security expenditure unresolved. The major powers continue to account for the vast majority of the outlays in the security sector. According to US Arms Control and Disarmament Agency data, the USA and the USSR accounted for approximately 60 per cent of world military expenditure in 1985; NATO and WTO accounted for nearly 80 per cent[8]. Nonetheless, domestic expenditure on the security forces is a drain on resources for many Third-World countries, and it is legitimate to ask if Third-World governments should not also participate in disarmament-development exercises (as called for in UNGA Resolution 3463). Indeed, the Committee of the Whole of the International Conference on the Relationship Between Disarmament and Development pointed out that,

"... The opportunity cost of military expenditures over the past 40 years has been and continues to be borne by both the developed and the developing countries ... The developing countries are doubly affected: (a) in proportion to the expenditure they incur themselves; and (b) because of the disturbing effect of military expenditure on the world economy[9]."

Finally, proposals that envisage funding development through savings generated by arms-control agreements (the 1978 French proposal, for example), have to come to terms with the fact that, to date, such agreements have probably not yet produced monetary savings. SALT I and II in the 1970s did not[10], and minor savings would tend to be applied to other portions of the "disarmer's" security budget. It has frequently been necessary to make trade-offs of this kind in order to win the approval of the military for the cuts in weapon system that are negotiated. This is a political problem which can only be resolved within each disarming nation if the sitting government has the power and the desire to do so.

It has been estimated that the recently ratified INF Treaty will produce $2-3 billion in savings each for the United States and the Soviet Union over a ten-year period. While it is not known how the United States might allocate these funds, it is not unreasonable to assume that they will be applied to other portions of the defence budget, particularly given the sizeable cuts that have recently occurred in US military spending. One can be sceptical of the strongly expressed desires of the Soviet government to apply savings produced by the INF Treaty to the civilian economy, in view of its proposal for a joint US-USSR flight to Mars. It has been estimated that such an undertaking would cost the United States approximately $2.5 billion over six years and it can be assumed that the cost to the USSR would be similar.

Implications of Disarmament for Development

The purpose of disarmament-development schemes is to expand the resources available to the Third World for development purposes while simultaneously reducing outlays in the military sector. Despite numerous proclamations recognizing the need for Third-World development and for an end to the arms race, consensus does not exist within the international community (or, indeed, within certain governments) that either increased development assistance or reduced military expenditure is desirable.

Let us assume, however, that the substantial political obstacles to disarmament-development proposals were overcome. The outcomes for Third World countries could be rather different depending on whether it were simply the industrialized countries of the "North" that disarmed, or if Third-World countries themselves reduced their outlays in the security sector. As suggested in Table 2, the positive developmental effects of trading expenditure in the security sector for expenditure on development are greatest for Third-World countries, when Third-World governments are among the disarmers.

Reductions by the North

Indeed, when the industrialized major powers are the sole disarmers, the developmental effects for the Third World may not be all that great. This is particularly true when the disarmers retain for their own use a large proportion of any money saved by arms-control agreements or reductions in military spending. It is even more true when such savings are not reallocated to the civil-sector economy.

Table 2. **Development Implications of Disarmament for North and South**

		Disarmer: North	Disarmer: South
Developer	North	Largest benefits for North	
	South	Development prospects for South if North disarms	Development prospects for South if South disarms Largest benefits for South

There are, nonetheless, some indirect benefits that might result from reductions in the military budgets. One potential positive effect would occur if reductions in security spending by the major powers reduced the latter's capacity to intervene in the Third World. Intervention has taken the form of supplying troops to take part in combat, providing military technicians to operate weapons, training local forces and financing the purchase of weapons.

If the major powers intervened less frequently in the affairs of Third World countries, this would, of course, not signal the end of any given conflict, nor even intervention, since Third-World countries also frequently intervene militarily in other Third-World countries. Reduced major-power intervention could, however, contribute to the peaceful resolution of conflicts, since many Third-World disputes are exacerbated by the intervention of third parties primarily interested in promoting their own interests, not those of the original parties to the conflict. In addition, active wars in the Third World are able to continue because of support from outside powers. If the capacity of the industrialized countries to provide this kind of assistance were to decline, some of the conflicts dependent on major-power support might be terminated.

It was pointed out in the previous section that arms-control agreements may not generate substantial savings if funds are re-directed to other portions of the security budget of the state in question. By the same token, it may not be realistic to assume that reductions in the arsenals of the industrialized countries will lead to less expenditure on the kinds of forces needed for intervention abroad. For example, to the extent that savings will arise from the recently ratified INF Treaty between the USA and the USSR, they will occur in the two countries' nuclear forces, not in their conventional forces.

Lower security spending in the industrialized countries which resulted in reductions in conventional forces could lead to a decline in arms sales to the Third World. Not only might the armed forces of the industrialized countries support fewer arms producers but the industrialized countries might reduce the security assistance that is necessary to enable many Third-World countries to procure arms from abroad.

At the same time, this situation might produce an increase in Third-World arms purchases as the arms producers remaining in the industrialized countries sought to recoup their losses by increasing sales in the Third World. Furthermore, arms producers in countries such as Israel, Brazil, South Africa, China and India might succeed in filling at least some of the gap so that Third-World countries would continue to devote resources to the military sector rather than to development.

The People's Republic of China has recently emerged as a major commercial arms supplier, ranking fourth in the world (after the USSR, the USA and France) between 1984 and 1987. China has capitalized on the Iran-Iraq war. Together these two countries purchased some 80 per cent of the arms delivered by China to the Third World between 1984 and 1987[11]. It is possible that the Chinese government, which has been seeking greater efficiency in its defence industrial sector, has sought to reduce its own procurement costs by selling abroad[12]. A substantial amount of the Chinese arms exports appear to be produced directly for export, which causes questions to arise concerning Chinese claims of extensive conversion of defenceindustrial sector.

Even if lower military expenditure in the industrialized countries did translate into a decline in procurement costs for at least some developing countries, it is important to remember that procurement of weapons accounts for a relatively small share of Third-World budgets. Operating costs, particularly personnel costs (salaries, wages, pensions) absorb an extremely large proportion of security expenditure in most Third-World countries[13]. Thus, even if the international trade in weapons did slow, this would in all likelihood not affect Third-World security expenditure as much as might be expected, unless the slowdown was substantial and of long duration. Whatever negative effects security expenditure has on development would be reduced substantially only under those circumstances.

Reductions by the South

Countries engaged in armed conflict, or involved in unresolved disputes which periodically flare into open warfare, are among the heaviest spenders on the security sector in the Third World. If it were possible to reduce Third World military expenditure, the capacity of these countries to wage war might be reduced and the peaceful resolution of conflicts might be encouraged. The peaceful resolution of conflicts would undermine one of the major justifications of security spending which is to protect countries against external threats. Fewer external threats would theoretically make possible further reductions in security expenditure.

It is important to remember, however, that while most governments justify spending in the security sector on the basis of external threats, it is often internal security considerations which play an important role in determining the level of security expenditure. For this reason, a diminuation in the level or quantity of external threats might not produce a correspondingly large decline in security expenditure. But to the extent that external security considerations were the driving force behind Third-World security budgets, regional security agreements would probably have to precede agreements on reductions in security expenditure.

If security spending could be reduced, it would increase the pool of financial, material and human resources available to Third-World governments for investment in development. Funds previously devoted to training troops and officers might be redirected towards civilian training/education programmes. Military medical facilities might be converted into civilian hospitals and clinics. Hard currency could be used to import machinery and technical assistance for civilian industry rather than weapons and other military-related goods and services. Technicians and R&D personnel employed in military factories could turn their attention to the pressing technological needs of the developing countries.

There is nothing automatic, however, about the transfer from the security sector to the civilian sector. Resources could easily be allocated to non-developmental purposes, or applied to projects which are unproductive from both a financial and a developmental perspective. The uses to which any money saved in the security sector would be put depends on decisions by governments. There is substantial evidence of poorly conceived investment decisions in developing countries in the recent past. Governments which give development a low priority are likely to invest less productively than those which seek to strengthen the economy in order to improve the welfare of all citizens. Nonetheless, the greatest opportunities for producing developmental dividends for the Third World from reductions in security expenditure arise directly out of reductions in Third World security budgets.

References

1. United Nations, International Conference on the Relationship Between Disarmament and Development, Report of the Committee of the Whole to the Conference: Draft Final Document, A/CONF.130/21, 9 September 1987, p. 3, para. 9-10 (hereafter, A/CONF.130/21). See also United Nations, General Assembly, Study on the Relationship between Disarmament and Development, Report of the Secretary-General, A/36/356, 5 October 1981.

2. United Nations, General Assembly, Consideration of Ways and Means of Releasing Additional Resources Through Disarmament Measures, For Development Purposes, in Particular in Favour of Developing Countries. A/CONF.130/PC/INF/8, 25 February 1986.

3. United Nations, Reduction of Military Budgets: International Reporting of Military Expenditures, Study Series 4, Sales no. E.81.19, New York: United Nations, 1981, p. 1.

4. Secretary-General of the United Nations, PO 131/2(4-1), 28 February 1979.

5. For more details on the UN Unified Reporting System, see Nicole Ball, Security and Economy in the Third World, Princeton, NJ: Princeton University Press, 1988, pp. 97-106.

6. Excerpts of Interview with Soviet Armed Forces Chief of Staff, New York Times, 30 October 1987.

7. World Bank, World Development Report 1986, New York: Oxford University Press, 1986, p. 220, Table 21: Official Development Assistance: Receipts.

8. United States Arms Control and Disarmament Agency, World Military Expenditure and Arms Transfers 1987, Publication 128, Washington, DC: March 1988, pp. 43, 47, 77, 81.

9. A/CONF.130/21, pp. 5-6, para. 26.

10. M. Leitenberg, "United States-Soviet Strategic Arms Control: The Decade of Détente, 1970-1980, And a Look Ahead", Arms Control 8:3, December 1987: pp. 213-64.

11. R.F. Grimmett, Trends in Conventional Arms Transfers to the Third World by Major Supplier, 1980-1987, 88-352 F. Washington, DC: Congressional Research Service, Library of Congress, 9 May 1988, pp. 2-5, 22-25, 46.

12. R.J. Latham, "People's Republic of China: The Restructuring of Defense-Industrial Policies", in Arms Production in Developing Countries, ed. J.E. Katz, Lexington, Mass: Lexington Book, 1984.

13. The composition of security expenditure in Third-World countries is discussed throughout Ball, Security and Economy.

Contributions of Science and Technology to the Alleviation of Underdevelopment

Bhalchandra Udgaonkar

The Widening Gap Between Rich and Poor

It has been increasingly realized during the last few decades that progress in science and technology has, for the first time in human history, made available to man the tools by means of which poverty, hunger, ignorance, destitution and underdevelopment could be banished from the surface of the earth. On the other hand, one is faced with the unfortunate situation wherein the number of people below the poverty-line, and the number of illiterates (the two groups have a very considerable overlap) show no signs of decrease over vast areas of the globe, and the gap between the poorer and richer countries, and between the poor and the rich within many countries, has actually been widening year after year.

It may be worthwhile to emphasize that the basic human needs: food, drinking water, shelter, clothing, fuel and fodder, education, health and work, represent an area of urgent social action, not merely on compassionate or humanitarian grounds, or because their neglect in the so-called Third World is likely to create an explosive situation for the richer part of the world, but also because their satisfaction is a minimum prerequisite for the restoration of human dignity to a large majority of the world's population, and for giving real meaning to democracy and human solidarity.

In an interdependent world, poverty is a global issue. As Javier Perez de Cuellar, Secretary-General of the United Nations, observed recently, "collective survival and prosperity demand that abundance be shared worldwide. Poverty anywhere is poverty everywhere"[1]. This was gradually and painfully realized in today's rich industrialized countries after the industrial revolution, leading to increased solidarity across social sectors and the emergence of a democratic framework. The same solidarity is now needed across nations at different stages of economic development, avoiding as far as possible the painful processes of transition that characterized the transformation in the rich industrialized countries of today. This may appear utopian, but it has to happen. Deep ethical questions are obviously involved. A veteran Pugwashite remarked to me last year that national boundaries had become irrelevant today, and that one must demolish these artifical barriers. When I asked if this referred only to the opening up of the markets of developing countries or also to the sharing of the planet's resources and the benefits of development, the conversation changed to a different topic.

A key question for some time has been whether international action, using all the tools of Science and Technology (S&T) at its disposal, can save the 1000 million people below the poverty line who are today essentially written off by the national and international systems. Sensitive people have even claimed that the condition of these 1000 million people is not better than that in the concentration camps of the Second World War, and that the more fortunate ones who do nothing about it are as culpable as those who did nothing when millions of their countrymen were dying in concentration camps.

In alleviating poverty and hunger, or in the fulfilment of basic needs, the role of S&T is small. Solutions exist. They can of course be improved by R&D, but they exist. What is more important is the role of socio-political commitment, of human solidarity within a country and across national boundaries[2]. For example, progress in agricultural science and technology has led to a situation in which there is enough food in the world to feed all the people and more. Yet, about a third of the world population is hungry. The problem of distribution, within a country and across national boundaries, is crucial.

Many communicable diseases disappeared from Europe long before modern medicines became available, as a result of the overall improvement in the living conditions of people. They continue, however, to take a heavy toll in the Third World. Child mortality rates and maternity deaths continue to be high in many Third World countries, though it is within human capacity to cut them down drastically. Examples can be multiplied.

Crisis of the World Economy

The World Economy has been going through a crisis. The collapse of commodity prices, the deterioration in terms of trade, and the increasing protectionism in the industrially-advanced rich countries have deprived the less developing countries of export earnings badly needed to sustain their development as well as to service their debts. Intolerable debt burdens have compelled many of them to adopt "adjustment programmes" at high social and political costs, including severe constraints on economic development. **Per capita** GNP has fallen in many countries. Stagnation of official development assistance and a relative fall in concessional finance has forced these countries to take high interest loans from commercial banks. The shift from low-interest long-period loans to high-interest short-period loans, and the volatile and fluctuating exchange rates, are other factors which have led to a net transfer of resources from the less developed countries to the developed countries. The less developed countries have thus become explicit financiers of the development of the developed countries, as also the shock absorber of the crises of the developed countries. Old colonialism has been replaced by a new form of exploitation: the less developed countries continue as subsidiary economies, as in the colonial era, supplying commodities and cheap labour. As soon as they try to diversify in the direction of industrial products, protectionism comes in the way.

It has become necessary to redefine what one means by terms like interdependence, international division of labour, reciprocity. The Brundtland Commission has recently drawn attention to the close links between poverty, underdevelopment, and environmental degradation. Underdevelopment increases the vulnerability of a country to a fall in commodity prices and adverse terms of trade, including protectionism in the rich countries; the decline in export earnings and abnormally high levels of real interest rates conspire to increase the debt burdens of the less developed countries to unbearable levels. This debt-trap not only leads to a drain of wealth from the Third World countries to the rich countries, but the less developed countries are forced to cut down on social welfare or poverty alleviation measures; to resort to environmentally unsound policies and production techniques; and thus to tax their natural resources beyond the limits of recovery, leading to environmental degradation. This too is a global issue since climatic changes in one part of the world affect others. As if this degradation of the environment were not enough, some rich countries not only transfer hazardous industrial operations to Third World countries, but they are reported to be dumping garbage and hazardous waste in these countries.

Contributions of Science and Technology

What are the contributions that S&T could make to improve this dismal situation? The role of science is in providing the knowledge base and the tools necessary for a systematic analysis of any complex problem and the working out of a solution - a strategy and an action plan. Technology provides the specific tools and skills to carry out efficiently the strategy and the task. Developing countries cover a wide spectrum, in terms of natural resource base, population, availability of trained manpower, and their capabilities in S&T and industry. Financial resources needed for development and for social services to ameliorate the living conditions are a big problem for most developing countries. Their expectations of substantial assistance from the rich developed countries have not materialized, though three Development Decades have gone by. "Aid" has often turned out to be mere pre-investment or, worse, a handle for influencing the development strategy, the economic policies, the political alignment and the decisions on equipment purchase in a direction preferred by the "aid-givers". The hope that disarmament may release vast financial resources for development has also not materialized.

Realistically, especially in the long run, resources for development will have to be generated within a country by judicious development of S&T capability in selected sectors - for example, for moving away from a single crop/mineral economy, or for gaining control of the processing of such natural resources into value-added products; for achieving self-sufficiency in food; for generating the capacity for autonomous decision making[3]. Going further - especially in the case of larger developing countries - or for groups of developing countries - for developing products, processes and production techniques in selected areas so as to become competitive in the international market.

As late starters in modern S&T, developing countries have had high expectations that international cooperation would develop their S&T capabilities. While international scientific co-operation for development has contributed to a certain extent to the building up of the S&T capability of developing countries, it has not risen to the expectations. There have been various hurdles, pitfalls and structural and other limitations arising from the current pattern of international relationships, of which S&T relationships are a part[4]. These problems led Pugwash, in 1973-74, to propose a draft Code of Conduct for Transfer of Technology, and some years later to prepare the Pugwash Guidelines for Scientific Co-operation for Development. Unfortunately, the Code of Conduct seems to have been buried, after years of negotiations in an intergovernmental forum. The Pugwash Guidelines were an input into the UN Conference on Science and Technology for Development (UNCSTD) in Vienna in 1979, and also greatly influenced the preparations leading to the Vienna Programme of Action. The North was, however, not interested in implementing the Programme of Action, though it was arrived at after intensive negotiations. I was talking about this with a distinguished scientist from the North, interested in global problems, and was taken aback by his cool remark to the effect that it was never expected to take off any way[5].

At UNCSTD, there was a special session on Science and the Future. It was proclaimed as a scientific session, to be distinguished from the other sessions which were dubbed as political. But even at this session of scientists, looking at the future, I found that some scientists were not willing to support in public what they were willing to support in private. For example, while they bemoaned the strong asymmetry in S&T efforts between the North and the South - the South accounting for a mere 3 per cent - they opposed the recommendation that the South's share of S&T efforts be raised to 20 per cent by the year 2000. This recommendation could not go through in a session of scientists.

There are other fora where the North-South dialogue has been unable to make any progress. The outcomes of UNCTAD-VII, the Uruguay Round, the negotiations on the Generalized Scheme for Trade Preferences (GSTP) have been disappointing. The major economic powers have not been interested in associating the developing countries meaningfully in discussion on global economic issues. The North-South dialogue needs to be purposefully revived. It should be responsive to the changed character and circumstances of the world economy. It has to include the reform of the monetary and financial system, through an international conference on money and finance for development and urgent measures for preserving and strengthening the multilateral trading system. There has also been a progressive erosion of the role of UN and its agencies. It is high time to reverse this retreat from multilateralism.

The nature of S&T transactions among countries, the role of transnational corporations, embargos on equipment and know-how, and the increasing restrictions on information flows, including scientific conferences, are other issues which call for serious attention. It is clear that a new Science and Technology Order has to be a part of a new World Order[6], whether one likes to call it the New International Economic Order or not. And scientists concerned about world affairs and peace have to be concerned about it.

We are far from being a world community of scientists. Scientists will begin to form such a community only if they are willing to apply their minds seriously to the question of what kind of **World Beyond War**, what kind of New World Order they stand for, transcending narrow national and ethnocentred interests. This kind of New Thinking has to gather momentum.

The Role of Pugwash

As mentioned earlier, Pugwash has from time to time been concerned about these problems. It is time these issues started becoming major concerns of Pugwash, as important as the East-West issues of disarmament, nuclear disarmament in particular, where there now appears to have been a major breakthrough[7]. Scientists are supposed to be a minority in league with the future. As an important segment of this community, Pugwash has to lead the way. And one has to move from sectoral technocratic approaches to the totality of the problem. An appeal from some Pugwashites to the members of the Pugwash Council at Gmunden in 1987 drew attention once again to the close relationship between armament/disarmament, underdevelopoment/development and peace and security, and emphasized that "such a reconceptualization of the message of Pugwash should involve a shift in the centre of gravity of peace and security from the exclusive reliance on military hardware reduction to the satisfaction of basic human needs as a crucial peace and security asset". It added "A second important point for ensuring peace is to stand against the ever more ruthless exploitation of the natural environment (e.g. the tropical rain forests) and the senseless waste of resources in the interest of small but powerful lobbies, denying to the Third World Countries their share of the Planet's treasures and allowing an obscene affluence for an already overfed privileged minority of conspicuous consumers in the rich countries in collusion with corrupt elites in some of the poor ones"; and it called for an explicit recognition that "the quest for stable and lasting peace is inextricably linked to the quest for social justice on the national, regional and global level. It could finally call for an overall Strategic Peace Initiative as opposed to the morally indefensible SDI"[8]. This appeal deserves the serious consideration of Pugwash.

References

1. Javier Perez de Cuellar, message to the 19th World Conference of the Society for Internationl Development, New Delhi, 28 March 1988.

2. B.M. Udgaonkar, "Action Programme on Research and Human Needs", in Federico Mayor (ed.) Scientific Research and Social Goals: Towards a New Development Model, Pergamon, 1982.

3. See, for example W.K. Chagula, B.T. Feld, A. Parthasarathi (eds.) Pugwash on Self-Reliance, Ankur, 1977.

4. P.J. Lavakare, A. Parthasarathi, B.M. Udgaonkor (eds.), International Scientific Coperation for Development: A Search for New Directions, New Delhi, Vikas, 1980.

5. See also, Abdus Salam, Ideals and Realities, World Scientific, 1987.

6. B.M. Udgaonkar, "Science and Technology in the New International Economic Order", in Science Policy for the Eighties and Beyond, M. Gibbons, P. Gummett, B.M. Udgaonkar (eds.), Longman, 1984.

7. Anatoly Gromyko and Martin E. Hellman, Breakthrough: Emerging New Thinking, 1988.

8. An appeal to members of the Pugwash Council by M. Thee and others, Gmunden, 5 September 1987.

Food, Population and Conflict in Africa

Roger Revelle

Some Statistical Data

Next to Asia, Africa is the largest continent, covering between a quarter and a fifth of all dry, ice-free, land on earth. Its mid-1980s population was 454 million people divided into 45 countries, ranging in population size from Nigeria, with at least 85 million people, to Swaziland, with 615 thousand. Outside the Sahara Desert, the average population density was about 25 people per square kilometre, in contrast to a density of 200 people per square kilometre in India, and 620 in Bangladesh.

Most of Africa's 45 countries are artificial creations established during the colonial period. Several of them, like Cameroun, contain numerous tribal groups, each with its own language. The countries' very existence depends upon the common language, either French, English, or Portuguese, imposed upon them by the European colonists.

Twelve of the 45 countries have been involved for years in civil or international wars. These include Angola, Chad, Ethiopia, Libya, Mauritania, Morocco, Mozambique, Namibia, Somalia, South Africa, Sudan, and Uganda. Together these 12 countries contain 153 million people, one-third of the population of the African continent. However, as Table 1 shows, overall military expenditures in Africa make up a smaller percentage of the continent's gross product than do military expenditures in developed countries as a fraction of their gross product. In 1982 only Angola, Ethiopia, Tanzania and Zimbabwe had ratios of military expenditures to gross national product higher than the average ratio for developed countries.

If modern, high-level agricultural technology were used, Africa has the agricultural potential to produce enough food for 10 billion people. Unfortunately, with the present continent-wide average rate of population growth of nearly 3 per cent per year, this population level would be reached in about 110 years, that is before the end of the 21st century. Table 2 shows demographic characteristics for the sixteen African countries listed in Table 1. Rates of natural increase vary from 2.1 per cent in Chad to 4.3 in Kenya, but most of the differences are due to differences in death rates. Exceptionally high birth rates of fifty or more per thousand are observed only in Kenya, Uganda, Zambia and Zimbabwe. The average African woman has six to eight children during her reproductive lifetime. With present rates of natural increase, the populations of most countries shown in Table 2 will double in 17 to 24 years. Low life expectancies and high infant mortalities in many countries at the present time suggest that slight improvements in health conditions would cause a fall in death rates and a significant rise in rates of natural increase during the next several decades.

In spite of its huge agricultural potential, African food production is at present not keeping up with population growth (Table 3). Food production **per capita** has declined over the past 20 years, and in several countries

Table 1

Country	Population	GNP	Military Expenditures	Military Imports	Education	Health
	millions	----------------billion of dollars				--------------
Sub-Saharan Africa	371	193	6.2	1.9	8.3	2.0
Angola	7.3	10.4	1.0	0.4	0.6	0.1
Burundi	4.4	1.2	0.05	0.02	0.04	0.01
Ethiopia	30.6	4.6	0.4	0.3	0.15	0.07
Chad	4.8	0.4	0.01	0	0.01	0.003
Congo	1.6	2.3	0.1	0	0.1	0.06
Guinea	5.3	1.7	0.07	0.005	0.05	----
Kenya	17.8	7.0	0.3	0.07	0.46	0.17
Mauritania	1.6	0.8	0.08	0.01	0.03	0.01
Mozambique	12.7	4.7	0.16	0.13	0.06	0.006
Somalia	6.1	2.1	0.12	0.14?	0.03	0.01
Sudan	20.0	9.4	0.3	0.18	0.4	0.03
Tanzania	19.8	5.3	0.32	0.02	0.28	0.09
Uganda	13.5	3.2	0.04	0.06?	0.06	0.02
Zaire	30.3	5.6	0.08	0.04	0.32	0.05
Zambia	6.1	3.9	0.12	0.02	0.21	0.14
Zimbabwe	7.8	6.4	0.43	0.09	0.57	0.16

Comparison With Developed Countries

Country	Population	GNP	Military Expenditures	Military Imports	Education	Health
All Developed	1071.3	9885.0	521.3	7.3	530.0	476.8
USA	232.3	3056.9	197	0.43	159.3	136.8
USSR	270.0	1563.0	170	0.68	77.7	48.0
Japan	118.4	1190.0	11.9	0.6	67.9	55.9
France	54.4	627	26.2	0.1	25.2	40.1
FRG	61.6	757	25.5	0.5	38.5	49.2
Sweden	8.3	115.4	3.8	0.09	10.4	10.4

Source: Ruth Sivard, World Military and Social Expenditures, Washington, D.C.: 1983.

Table 2

Country	Crude Birth-rate per 1000	Crude Death-rate per 1000	Percentage Rate of Natural Increase	Total Fer-tility Rate	Male Life Expectancy Years	Infant Mortality per 1000 Live Births
Angola	49	22	2.7	6.5	42	165
Burundi	47	19	2.8	6.5	45	123
Ethiopia	47	18	2.9	6.5	45	122
Chad	42	21	2.1	5.5	42	161
Congo	43	10	3.3	5.7	59	68
Guinea	49	27	2.2	6.5	37	190
Kenya	55	12	4.3	8.0	55	77
Mauritania	43	19	2.4	6.0	43	132
Mozambique	49	16	3.3	6.5	49	105
Somalia	48	25	2.3	6.5	38	184
Sudan	45	18	2.7	6.6	46	119
Tanzania	47	15	3.2	6.5	51	98
Uganda	50	19	3.1	7.0	46	120
Zaire	46	16	3.0	6.3	49	106
Zambia	50	16	3.4	6.8	49	105
Zimbabwe	54	12	4.2	8.0	54	83
Comparison With Developed Countries						
USA	16	9	0.7	1.8	71	11
USSR	19	10	0.9	2.4	65	--
Japan	13	7	0.6	1.7	74	7
France	14	11	0.3	1.8	71	10
FRG	10	12	-0.2	1.4	70	12
Sweden	11	11	0.0	1.7	75	7

Source: IBRD World Development Report 1984.

Table 3

Social and Economic Characteristics of Some African Countries

Country	Urban Population Percentage 1982	Urban Growth Rate Percentage 1982	Percentage of Labour Force in Agriculture 1980	Index of Food Production Per Caput 1980-82*	Daily Calorie Supply 1981	Percentage of Requirement 1981
Angola	22	5.8	59	77	2096	83
Burundi	2	2.5	84	96	2152	95
Ethiopia	15	5.6	80	82	1758	76
Chad	19	6.4	85	95	1818	76
Congo	46	4.4	34	81	2199	94
Guinea	20	5.2	82	89	1877	75
Kenya	15	7.3	78	88	2056	88
Mauritania	26	8.1	69	73	2082	97
Mozambique	9	8.1	66	68	1881	76
Somalia	32	5.4	82	60	2119	100
Sudan	23	5.8	78	87	2406	99
Tanzania	13	8.5	83	88	1985	83
Uganda	9	3.4	83	86	1778	80
Zaire	38	7.6	75	87	2135	94
Zambia	45	6.5	67	87	2094	93
Zimbabwe	24	6.0	60	87	2025	90
Comparison With Developed Countries						
USA	78	1.5	2	119	3647	138
USSR	63	1.8	14	101	3328	130
Japan	78	1.8	12	91	2740	117
France	79	1.4	8	121	3360	133
FRG	85	0.5	4	118	3538	133
Sweden	88	1.0	5	119	3196	119

* 1969-71 = 100

Source: IBRD World Development Report

food production has actually diminished during this period. This would not be too serious if production of export crops such as cacao, cotton, palm oil and sisal, that could be traded for food with other countries, had increased sufficiently. But as Table 3 shows, overall food supplies, even including trade imports and food aid, have not been sufficient to provide a healthy diet for poor people in many countries.

Reasons for Poor Performance of Agriculture

Table 3 shows the overwhelming predominance of agriculture in African countries, where on the average more than two-thirds of the labour force is engaged in farming. Urbanization is progressing moderately rapidly in some countries, such as Zaire, Zambia, Zimbabwe and Kenya. But in general, most Africans are rural and depend economically on agriculture. One of the reasons for the continuance of pervasive African poverty stems from the poor performance of agriculture. What are the reasons for this?

1. Inadequate water supplies. The Sahelian region, just south of the Sahara Desert, has suffered from an almost continuous drought for the past fifteen years. Droughts have also been prevalent in southern Africa. Large-scale irrigation has been successful in only a few countries. (Egypt, Sudan, and Zimbabwe).

2. Poor Soils. In some areas, vertisols (sticky, cracking, crusty, "black cotton soils") prevail. In many other areas a brick-like lateritic hardpan exists a few inches below the surface. Elsewhere many soils are acid and highly leached, with low base exchange and water- and fertilizer-holding capacity, and high alumina and ferric iron content.

3. Rapid soil erosion. When lands are cleared of vegetation, rapid soil erosion takes place, both from violent rainfall and from wind. A one-centimetre layer of soil may be removed in five to ten years. In one of the processes of desertification arable lands may be smothered by wind-borne sands.

4. High incidence of debilitating and often fatal diseases. Besides the usual diseases of tropical or semi-tropical countries, (malaria; leprosy; tuberculosis; diarrheal and enteric diseases; acute respiratory infections, including measles, diphtheria and pertussis; arbovirus infections, including yellow fever; schistosomiasis; leishmaniasis; and filariasis), two peculiarly African diseases are devastatingly severe over much of tropical Africa. These are African trypanosomiasis, a protozoan disease, carried by tsetse flies, of cattle and human beings (in the latter it is the usually fatal "African sleeping sickness"), and onchocerciasis, or "river blindness," a filarian worm disease, carried by black flies.

5. Poor transportation. Most African countries have an inadequate or partially decayed road system.

6. Low farm prices for food crops. Often such low prices that the farmers cannot afford to grow produce for sale. Governments maintain low prices to placate urban populations, and at the same time impose heavy taxes on agriculture, which is usually the main component of national production.

7. Inadequate or unavailable farm credit. Crop loans for food crops are virtually non-existent, and the communal land tenure system precludes the use of farm land as security for a loan. (It also tends to inhibit land improvements by individual farmers).

8. Inadequate farm inputs. Because of low farm prices and unavailability of credit, most African small farmers use little or no fertilizer, pesticides, herbicides, or improved crop varieties.

9. Most farmers are women. Because of their low status in society, women have difficulty profiting from their own farm production, and therefore tend to grow just enough for their families.

10. Little experience with irrigation. In general, African farmers, except in Egypt and the Sudan, are unfamiliar with modern irrigation technologies.

11. Inadequate extension services. Extension workers are nearly all males, who do not communicate with women farmers. Moreover, the large number of small farmers is usually beyond the capacity of even competent extension workers.

12. Farm labour shortage. In the virtual absence of mechanisation, and of animal traction power in the tsetse fly areas, farm labour is insufficient during peak farming periods, such as times of planting and harvesting.

13. Inadequate agricultural research. Despite the existence of several international research centres in Africa, under the aegis of the Consultative Group on International Agricultural Research, sponsored by the World Bank, the Rockefeller and Ford Foundations, USAID, and other national aid agencies, crop genetic research has failed to produce high productivity crop varieties which are adapted to African climatic and ecologic conditions. This is in marked contrast to the high yielding varieties of wheat and rice that are the basis of the Green Revolution in Asia and parts of Latin America.

14. Gross disparity between experiment station and farm yields. In Ivory Coast, for example, yields of casava tubers in the agricultural research station are 30 tonnes per hectare, whereas the average farmer's yield are 3 tonnes per hectare. This disparity may be largely due to a combination of several of the factors listed above, but there may also be other, less evident reasons.

What is to be done?

1. Long-range agricultural research on sustainable soil and water management, development of high-yielding, environmentally adapted, pest-resistant, crop and tree varieties, and of improved breeds of cattle and other livestock. Modern techniques of tissue culture, recombinant DNA technology, storage and use of plant germplasm, and use of artificial insemination and embryo transplants in livestock improvement, are among the technologies which should be developed and applied.

2. Development of post-harvest and food technology. Inordinate losses of foods could be avoided by better crop storage facilities and improved technologies for food processing.

3. Improved agricultural extension services including particularly extension services for women farmers.

4. Expanded higher educational programme for agronomists, soil scientists, agricultural geneticists, hydrological engineers, and other specialists in agriculture and animal husbandry, particularly competent research workers.

5. Greatly increases cooperative planning, coordination and integration of international aid programmes in agriculture and related fields.

6. Improvements in the education and status of women, including creation of price and commodity incentives for women farmers to increase their production.

7. Reduction of population growth rates. This depends in large part on item 6 above, but it will also require a marked change of attitude by both men and women.

Zimbabwe's Agricultural Renaissance

While food shortages continue to threaten many countries south of the Sahara, Zimbabwe stands out as an exception. By contrast to the situation in 1980-82 (Table 3), both commerical agricultural production and peasant agriculture are now flourishing. Today, Zimbabwe is not only self-sufficient in its stable food, maize, but exports much of its surplus to other African countries. In 1987, the country had its biggest maize harvest ever, resulting in a surplus of 800 000 metric tonnes. As a result of its success, Zimbabwe has been able to give food to Mozambique (25 000 metric tonnes of maize), Tanzania (25 000 metric tonnes of maize) and Ethiopia (12 500 metric tonnes of maize and 12 500 metric tonnes of sorghum) to help alleviate shortages in those countries.

The contribution of peasant farmers to Zimbabwe's total agricultural production rose from ten per cent in 1979 to 64 per cent in 1987. Today, agricultural production among Zimbabwe's small farmers is among the highest of all peasant farmers in Africa. Some of the keys to this success have been:

1. Equal rights for women. Women in Zimbabwe do two-thirds of the agricultural work. Traditionally not allowed to own land, they were passed over when candidates were chosen for loans or training courses. In 1981 the Zimbabwean Parliament granted them equal rights with men in land ownership, agricultural training and credit.

2. Guarantees of high producer prices. To boost production by small farmers, the government has set prices high enough to guarantee them a profit and provide the incentive to grow more.

3. Infrastructure. A network of distribution centres makes seeds, fertilizers, pesticides and tools readily available. New grain storage depots and crop collection points enable farmers to market their produce without paying high transport costs. Storage losses are also kept low - less than 1 per cent, compared with levels of 15 per cent in some other African countries. And the number of agricultural extension workers has been increased, to help small farmers adopt modern methods.

4. Land reform. As part of the Government's programme of support of black peasant farmers, 37 000 families have been provided with a total of 2.7 million hectares of land.

5. Credit programmes. In 1985, the Government loaned about $35 million that went to fewer than 4500 small farmers in 1979, the year before independence. Repayment rates have been high.

6. Encouragement of local associations. Robert Mugabe's Government encourages and maintains open lines of communication with peasant groups. The

National Farmers Association of Zimbabwe comprises 400 000 families working small and communal farms.

Racial and Ethnic Peace

Mugabe's imaginative policy of reconciling bitter antagonisms across the colour line in a war-ravaged land succeeded far beyond the expectations of most observers. He sought to create a sense of security for both the winners and the losers of the bloody conflict in which the state of Zimbabwe was born.

A white minister of Agriculture was appointed in a successful effort to secure the confidence of the white commerical producers. During the period in which output of peasant farms increased ten-fold, white-owned commercial farms also prospered, increasing their output three-fold.

President Mugabe has also stressed international cooperation in African agriculture[1]:

> "Cooperation between countries in the fields of food production, trade and other economic activities is essential if we are to exploit fully the complementarities which exist between our countries and maximize regional self-reliance in food production. Regional organizations can be established with the support of the international community and would be instrumental in ending hunger in Africa and achieving self-sufficiency on the continent".

Reference

1. R. Mugabe, A Shift in the Wind, no. 22, April 1986.

The Role of Health Care in Alleviating Underdevelopment

Andrew Haines

Economic and social factors are clearly major determinants of health, nevertheless, health care can have a significant role in alleviating the effects of underdevelopment. Simple economic measures capable of preventing large numbers of deaths in the Third World, particularly amongst children, have made major improvements in health a possibility. However, the inappropriate delivery of health care, leading, for example, to excessive concentration of limited funds on relatively high technology hospitals in major cities, can actually inhibit advances in health.

Child Health

Despite the difficult economic climate in many Third World countries the promotion of low cost means of treating common health problems of childhood is already resulting in the saving of nearly 40 000 childrens' lives each week[1]. Nevertheless, more than a quarter of a million young children still die each week in the Third World. The major causes of death are infections and prolonged undernutrition (Table 1).

Table 1. Annual deaths of children under five in developing countries (1987) (Adapted from UNICEF 1988)

Cause	Numbers (million)	Comments
diarrhoeal diseases	5	3.5 million could be prevented or treated using oral rehydration therapy
acute respiratory infections	2.9	0.6 million deaths from whooping cough could be prevented by immunization, many others could be treated with antibiotics
measles	1.9	Could be prevented by one vaccination
malaria	1	Could be greatly reduced by low cost drugs
tetanus (neonatal)	0.8	Could be avoided by immunization of expectant mother
other	2.4	Many could be avoided by antenatal care, breast feeding and nutrition education

Although in the Table a single cause of death has been given, many children die of multiple causes and malnutrition is a contributory cause in approximately one third of all child deaths.

Half of all illness in children is associated with poor hygiene and lack of safe water. Although the provision of piped water may be expensive, simple measures, such as the use of latrines, can have a worthwhile impact in reducing the burden of disease. The timing of births is also a major factor contributing perhaps 25 per cent of current childhood deaths. Pregnancies which are too close together, or to mothers younger than 18 or older than 35, increase the risks for both mother and child. Infant mortality ranges from 6 per 1000 live births (Sweden) to 198 per 1000 (Sierra Leone).

Diarrhoea is the leading cause of child death. Oral rehydration therapy (ORT) can be initiated using simple mixtures of sugar, salt and water. If the diarrhoea persits specially formulated oral hydration salts (ORS) can be given. These cost only a few cents and can be administered by parents with some instruction from a health worker. Despite the two decades which have elapsed since the discovery of oral rehydration therapy, in many parts of the world only a small proportion of episodes of diarrhoea are treated effectively in this way. In Africa the proportion has been estimated as less than 10 per cent; globally (excluding China) it amounts to around 18 per cent. The World Health Organization (WHO) has estimated that only about 25 per cent of health facilities stock ORS, and less than 10 per cent of health workers have been adequately trained in its use. Many doctors and health workers still prescribe anti-diarrhoeal drugs which may actually do harm.

In the 1980s three vaccine-preventable infections - measles, whooping cough and tetanus - have killed around 25 million children, approximately equivalent to the entire population under fives in the USA or Western Europe. About half a billion doses of vaccines were given in 1986 and the percentage of immunization coverage is increased by about 7 per cent per year, but only around 40 per cent of children in the Third World are covered by the age of one. The main problem is not a lack of vaccines or "coldchain" equipment for keeping vaccines at low temperatures, but of informing parents and developing the infrastructure to undertake reliable vaccination programmes.

The cost of immunization in the Third World has been calculated as approximately $500 million - approximately equivalent to the cost of ten advanced fighter aircraft. Every two seconds a child dies and another child is permanently and severely disabled by preventable diseases whilst $50 000 is spent on arms. The total expenditure on arms in one year is approximately equivalent to the total gross national product of the countries in which the poorest half of the world's population lives[2].

Other Health Problems

Around 100 million people have no shelter whatsoever, and 1.3 billion do not have safe water to drink. Around 500 million suffer from iron deficiency anaemia due to dietary deficiency or intestinal parasites such as hookworm. Vitamin A deficiency is a common cause of blindness which is easily preventable. Tuberculosis is still a major cause of death in many countries, despite well proven treatment regimes with very high cure rates. Life expectancy ranges from 77 years (Sweden, Japan, Ireland) to 37 years (Afghanistan)[3].

It has been suggested that the spread of AIDS could be reduced by perhaps 80-90 per cent over the next ten years if the public were informed and took appropriate steps to avoid infection. The disease is likely to infect 5-30 millions over the next decade. The spread of AIDS has major implications for health care in most countries of the world. In Africa the sex ratio for AIDS sufferers is close to unity, thus many young women of child bearing age are affected. Nearly half of their babies also become victims. Thus in Zambia it has been estimated that 6000 babies may have AIDS, over 10 times the number of those in the USA where AIDS is still a disease which predominantly affects males. There is a danger that AIDS sufferers may overwhelm the health facilities in many countries, particularly those which are already under great pressure.

Medical Service Taining

There are now around 2 million doctors and 6 million nurses in the Third World. The numbers have doubled over the last fifteen years or so. Many millions of community health workers have been trained to work at grass roots level. Unfortunately, in many countries the opportunities for professionals to give information about immunization, breast feeding, oral rehydration and birth spacing is frequently lost because of inappropriate training, or out-of-date knowledge. Nevertheless, many medical societies and associations are putting their weight behind efforts to increase public knowledge of these basic methods of promoting health. Despite the fact that hospitals can address only a small proportion of the population's health needs, the great majority of medical schools and other training institutions are hospital-dominated to an excessive degree. In recent years, however, there has been increasing interest in a more community oriented approach to medical education. Several medical schools around the world have developed "innovative tracks" in education which emphasize problem solving skills and community based learning[4].

The development of non-hospital based primary (first contact) care is an essential cost effective approach which emphasizes the crucial role of community personnel in delivery care. In Tanzania, for instance, for the cost of training one doctor (£14 700), 17 rural medical assistants or 35 medical aides could be trained[5]. Effective primary care is based on several principles; (a) education about diseases, health problems and their control; (b) safe water and basic sanitation; (c) maternal and child care, including family planing; (d) immunization; (e) appropriate treatment of common health problems; (f) provision of essential drugs.

The activity of health promotion is not the exclusive role of health services. In order to be effective, many of the simple measures of health promotion both in developed and Third World countries depend on wide social involvement. The support of teachers, religious and political leaders, mass media, government agencies, voluntary organizations, trade unions and other organizations, is of key importance.

Health 2000

The Thirtieth World Health Assembly resolved in May 1977 that "the main social target of governments and WHO in the coming decade should be the attainment by all citizens of the world by the year 2000 of the level of health that will permit them to lead a socially and economically productive life". This proposal implies a much greater level of equity between people and its major thrust is on health promotion and prevention. A well motivated and actively participating community is a key element in the strategy, and it is acknowledged that health authorities can deal only with part of the problems to be solved, and that multi-sectoral cooperation is the only way to effectively ensure a healthy society. It acknowledges that the focus of the health care system must be primary care with services being provided as close as possible to where people live and work, with ready accessibility. Specific targets have been agreed in some regions; for instance, the member states of the European region have agreed on a series of 38 targets which in many cases provide specific guidelines on priorities, such as the elimination of specific diseases which can be prevented by immunization, improvements in life expectancy, increase in the percentage of disabled persons engaged in regular occupational activities, reduction in coronary disease, stroke, and cancer of specific sites, and the development of mechanisms for the assessment of the quality of health care.

More recently WHO has been promoting the concept of the district health system in which primary care services together with a district hospital form part of the integrated whole, to provide more comprehensive care to a defined geographical area within a country.

Threats to Health

In the 1980s more than 70 governments have had to adopt economic "adjustment" policies in an attempt to reduce balance of payments deficits. Many of these policies involve cut-backs on health and social services and on subsidies for staple foods. Around 29 per cent of the population of the Third World live in countries with zero or negative growth between 1979 and 1985. A special UNICEF report published in 1987 - "Adjustment with a human face" set out specific alternatives to current strategies. This document proposed that the effect of these policies on the most vulnerable sections of society, particularly children, should be assessed in advance, and that services and subsidies which are essential to survival should be strengthened rather than cut. Several countries have responded by reducing funding on expensive hospital building programmes whilst expanding primary care facilities.

In many countries military expenditure is considerably in excess of expenditure on health. The difference is even more noticable in many third world countries, amounting on average to a fourfold difference between military and health expenditure[3]. Although it cannot be assumed that reductions of expenditure on defence could necessarily be transferred to health, it is nevertheless salutory to note that the total budget for the World Health Organization for 1988/9 is $634 million which is equivalent to the cost of around 6 hours of world military expenditure.

Cooperation for Health

International cooperation for health is vital in alleviating the effects of underdevelopment. As well as UNICEF and WHO there are a wide range of non-governmental organizations dedicated to improving health around the world. Recently, for example, Action in International Medicine, an organization which brings together Colleges and Academies of medicine from many countries, has been formed to involve the medical and allied professions in improving health care and medical training by cooperative ventures in the Third World[6]. Its initial plans include implementations of WHO proposals for district health systems in several locations around the world.

International collaboration for health can have an impact in preventing deaths and morbidity in the Third World. Diversion of even a small proportion of military expenditure to health would greatly accelerate progress in alleviating some of the effects of underdevelopment.

References

1. UNICEF, The State of the World's Children, Oxford University Press, 1988.

2. Sidel, V.W., "Destruction before Detonation" The impact of the Arms Race on Health and Health Care, Lancet 1985, 2; pp. 1287-9.

3. Sivard, R.L. "World Military and Social Expenditures", 1987-88, Washington D.C., World Priorities Inc., 1987.

4. Kantrowitz, M., Kaufman, A., Mennin, S., Fulop, T. & Guilbert, J-J. Innovative tracks at established institutions for the education of health personnel, Geneva, World Health Organization, 1987.

5. Sanders, D., The Struggle for Health, Medicine and the policies of underdevelopment, 1986, London, Macmillan.

6. Wolstenholme G., "Action in International Medicine", Information Bulletin No. 1, Royal College of Physicians, 11 St Andrews Place, London, NW1 4LE.

Appendix

Global Problems and Common Security

Statement of the Pugwash Council

The 38th Pugwash Conference on Science and World Affairs met in Dagomys, near Sochi, from 29 August to 3 September 1988. The participants comprised 189 scientists, other scholars and public figures - as well as 29 students - from 41 countries and 4 observers from international organizations.

Greetings from General Secretary Mikhail Gorbachev, and his endorsement of Pugwash's efforts, were conveyed to the participants at the opening session of the Conference. Messages of support were also received from several other heads of state, from the Secretary General of the United Nations, and from a variety of other international organizations. The participants were greeted at a reception in Moscow following the meeting by the President of the Soviet Academy of Sciences, Academician Gurij I. Marchuk. The Pugwash Council wishes to express its sincere appreciation to these officials and to all of our Soviet hosts. Special thanks are due to the Chairman of the Soviet Pugwash Committee, Academician Anatoly A. Gromyko, Professor Sergei P. Kapitza, and Dr. Vladimir P. Pavlichenko - as well as to the staffs of the Dagomys Tourist Complex, the Central House of Tourists in Moscow, and the African Institute of the Soviet Academy of Sciences. The support of the Soviet Peace Foundation and particularly its First Vice Chairman, Vladimir P. Maslin, is also gratefully acknowledged. We also thank the John and Catherine MacArthur Foundation, the W. Alton-Jones Foundation and UNESCO for financial assistance for this Conference.

This 38th Pugwash Conference took place in an atmosphere of optimism but also of urgency. The optimism derives from the many indications, since the 37th Conference a year ago, that a genuine transformation of international relations for the better is not only possible but actually has begun. The sense of urgency arises from recognition of the necessity of energetic action to solidify and build upon these gains, and to begin to bolster the so far inadequate international response to the threats to security and well-being associated with problems of environment and development.

Working Groups that met repeatedly throughout the week-long Conference treated (1) strategic nuclear disarmament and preventing the weaponization of space; (2) European security; (3) chemical weapons; (4) military research and development; (5) influence of global environmental problems on international relations; and (6) alleviating underdevelopment, poverty, and hunger. Plenary sessions treated "Cooperative Initiatives in Science, Technology and Development" and "The Social Responsibilities of Scientists".

A theme that emerged from these deliberations with particular force and clarity was the need for more widespread appreciation of the ways in which growing problems of environmental disruption, underdevelopment in the South, and overconsumption in the North relate to each other, link the interests of all the world's peoples, and interact with the problem of avoiding war that has long been the core concern of Pugwash. Convinced that the importance and timeliness of this conception of the interconnectedness of the threats to human well-being justifies special efforts to communicate it, we are taking the unusual step of issuing a separate "Dagomys Declaration of the Pugwash Council" on this topic, in parallel with the present statement on the Conference as a whole.

Both statements are the responsibility of the Pugwash Council alone. They do not attempt to cover all the topics discussed at the meeting, and they do not necessarily reflect the views of all the attenders. Participants in Pugwash Conferences take part as individuals, not as representatives of their governments or institutions, and they hold a wide range of views. Summaries of the discussions in the Working Groups of the Conference have been prepared by the Rapporteurs and Conveners of these groups and will become part of the Proceedings of the Conference to be published by the Pugwash office.

The evidence that a positive transformation of international relations is actually underway includes: the successful conclusion of the Intermediate Nuclear Forces Treaty with its unprecedented verification measures; the commencement of serious negotiations in pursuit of deep reductions in strategic nuclear forces; a marked increase in the extent of communication and cooperation between the countries of Eastern and Western Europe; the strong expressions of interest by the leadership of the Warsaw Treaty Organization, and increasing interest on the NATO side as well, in reducing and restructuring conventional forces along more strictly defensive lines; some significant steps toward military disengagement of the major powers from the Third World; some important successes, at last, for United Nations conflict-resolution efforts; and a foundation, in the Montreal Protocol on Ozone Depletion, the ICSU's "Global Change Program", and the "Tropical Forest Action Plan", for increased international cooperation in understanding and coping with large-scale environmental problems.

On the other hand, what has been accomplished so far is barely a beginning. There still is more talk about disarmament than there is action, and deployments of new weapons systems are proceeding at a rapid rate. The huge military forces opposing each other across borders in North and South alike suck human and material resources away from compelling human needs, while feeding fears and tensions and obstructing cooperation on common interests. Actual armed conflicts rage or smolder in several regions of the South, multiplying the poverty that is so widespread in that part of the world. Military research and development claims topically 30 per cent or more of the scientific talent and funds in weapons oriented industrial nations, slowing the rate of discoveries relevant to solving problems of energy, environment, transportation, health, and efficient industrial and agricultural production. The growth of many of the most threatening environmental problems continues to far outpace the limited controls imposed until now. Financial and technological assistance from the North to the South remains a pathetically tiny trickle into an ocean of need, and besides is often misdirected. And most decision-makers and members of the public seem unaware of the linkages among these phenomena.

Desperately needed, therefore, is a program of education and action to consolidate, connect, and accelerate the positive elements mentioned earlier - and some others - yielding an approach to the linked problems of militarism, environment, and underdevelopment that matches the scale and comprehensiveness of the threats. The goals of this effort should be no less than:

- a nuclear weapon-free world;

- the creation of a framework of international economic, technological, cultural, and humanitarian interaction and cooperation with which the use of nuclear weapons and other weapons of mass destruction becomes unthinkable;

- the full demilitarization of international relations;

- the closing of the prosperity gap between North and South; and

- a transition to technologies of energy supply, agriculture, and industrial production that are both environmentally sustainable and resource-efficient.

Among the many ideas discussed at the Conference for moving decisively toward these goals, we summarize in what follows only a subset to which we wish to give particular emphasis.

Achieving deep cuts in strategic nuclear forces depends on protecting the ABM treaty against both piecemeal erosion and radical reinterpretation. To accomplish this, both sides should recommit themselves to the ABM treaty in its original ("narrow") interpretation, accompanied by clarification, in the Standing Consultative Commission, of certain key terms (e.g., "components", "development", "testing") and by a ban on development, testing, and deployment, testing, and deployment of antisatellite (ASAT) weapons. The actual prescriptions adopted for cutting strategic forces should aim not so much at reducing numbers alone as at reducing characteristics of the forces that could promote crises instability.

Besides the prospect of deployment of strategic defences, the second major obstacle to achieving deep reduction in strategic nuclear weaponry is the potential exclusion, from the reductions scheme, of nuclear-armed sea-launched cruise missiles SLCMs. These difficult-to-verify systems would best be banned altogether. Making such a ban adequately verifiable might require either banning conventionally armed as well as nuclear armed SLCMs, or eliminating all nuclear warheads from surface naval vessels and allowing access to such vessels by inspectors.

A comprehensive ban on testing nuclear explosives (CTB) is needed to help stop dangerous trends in the qualitative evolution of nuclear weapons systems and to signify the commitment of nuclear weapons state signers of the Non-Proliferation Treaty (NPT) to fulfill their obligations under article VI of that treaty. We therefore commend the six-state initiative to convene an amendment of the signatories of the partial test ban treaty as a route to achievement of a CTB. Even under an arms-control regime that included deep cuts in strategic forces and a CTB, the maintenance of any nuclear arsenals whatever, and of doctrines of nuclear deterrence, would pose dangers of nuclear-weapons use as well as logical inconsistencies for non-proliferation policy. This circumstance underlines the need to move as rapidly as possible toward an international relations regime in which nuclear weapons could be eliminated altogether.

The reductions in nuclear forces that could soon be achieved must not become an excuse for needless, dangerous, and costly expansions of conventional forces, but rather should be the occasion for seeking ways to increase stability and confidence at much lower levels of conventional armaments. The key to accomplishing this is tying conventional force reductions and restructuring together in a way that removes the most threatening offensive elements from each side. Increasingly detailed and plausible prescriptions for doing so are being put forward by analysts and leaders on both sides, and it is clear that a beginning by way of an initial comprehensive data exchange on conventional forces, backed up by inspections, is within early reach.

Increasing security in Europe can be most expeditiously accomplished by a programme not limited to reductions and restructuring of military forces, but including in addition a wide range of confidence- and security-building measures extending from the military arena into the political, economic, cultural, and environmental spheres. Strengthening of the "Stockholm I" CBMs and intensive exchanges of views on doctrines are examples on the military side; expanded economic cooperation and integration, increased collaboration on the solution of regional environmental problems, and enlarged cultural exchanges are examples on the non-military side.

The increasingly dangerous situation in chemical weaponry demands that all states possessing chemical weapons swiftly and unilaterally renounce the production, stockpiling, dissemination, and use of these weapons and begin their destruction immediately. Such acts would in no way substitute for a rapid conclusion of a Chemical Weapons Convention totally banning these weapons. They would instead help to remove obstacles still in the way of this treaty, just as the 1969 decision of the Nixon Administration to unilaterally renounce biological weapons helped to conclude the Biological Weapons Convention. The failure of the international community to forcefully condemn the illegal and outrageous use of chemical weapons by Iraq helped to undermine the existing international regime as underlined in the 1925 Geneva Protocol and may have encouraged further proliferation of chemical weapons. This appalling set of events, and other reported uses of chemical weapons, underlines the need to move rapidly to a comprehensive CW convention.

Much military research and development is counterproductive because it tends to increase destructiveness of weaponry, provoke countermeasures, and undermine arms-control agreements, as well as diverting material and human resources from meeting real human needs. More systematic assessment of long-term consequences of military R&D might help avoid those R&D directions most likely to decrease security, and expanded studies of the conversion of military R&D capabilities to civilian tasks could reduce the economic dislocations of desirable overall cuts in military R&D. That relatively small component of military R&D that supports progress in developing optimum arms-reduction schemes and in enhancing verification capabilities should be expanded.

Reducing many of the most serious regional and global environmental threats depends on achieving increased efficiency of energy use, both in developing and industrialized countries. Increased energy efficiency - doing more with less - is the fastest, cheapest, and most dependable way to increase the quantity of energy services available to support economic growth and basic human needs while reducing deforestation for fuelwood and shrinking emissions of climate-threatening carbon dioxide and acid rain-producing oxides of sulfur and nitrogen from combustion of fossil fuels.

The next-best energy option with respect to reducing environmental threats would be expanded use of selected renewable energy sources - including photo-voltic cells, solar hydrogen production, and clean-burning approaches to use of biomass fuels. But bringing these possibilities to fruition soon in economically affordable as well as environmentally attractive form will require major increases in national and international programmes of renewable-energy research, development, commercialization, and diffusion. It is possible that improved nuclear-energy technologies - including, in the long longer term, nuclear fusion - can help reduce the fossil-fuel-linked hazards of energy supply, but this may require even larger R&D efforts than for the renewables before it becomes clear whether a satisfactory combination of environmental, safety, and economic characteristics is achievable.

To assure that the benefits of development reach those who need it most, increased emphasis in aid and trade policy must be placed on meeting basic human needs with the help of greatly expanded North-South and South-South cooperation. Essential conditions for sustainable development include improved terms of trade for raw materials and basic commodities imported from the South, rescheduling of the crippling burden of foreign debt, and increased availability of loans and credits. Transfer of advanced technology from North to South combined with national economic policies to increase the economic-efficiency of resource use can help to solve many development problems and promote self-reliance. The deadlock in the United Nations over adopting a code of conduct for the transfer of technology must be broken.

Alleviating poverty and hunger must remain a central issue on the international agenda. Priority should be given to the nurture of children. Low-cost and participatory projects to provide adequate nutrition are essential.

Infrastructure development, population planning, and the conversion of military units to civil tasks can contribute importantly to development efforts. The poor and the poor countries should have better access to political decision-making processes at national and international levels, helping them to develop self-reliance and confidence in their future. Increased educational efforts are required in both developing and industrialized countries to improve public understanding of the current situation and to increase the resilience of developing country economies to the range of stresses they confront. Success in these tasks will reduce the pressure on environmental systems, will decrease regional tensions, and lower the probability of future conflicts.

Dagomys Declaration of the Pugwash Council "Ensuring the Survival of Civilization"

We live in an interdependent world of increasing risks. Thirty-three years ago, the Russell-Einstein Manifesto warned humanity that our survival is imperiled by the risk of nuclear war. The familiar challenges identified in that Manifesto and the 1982 Warsaw Declaration of Nobel Laureates remain as important as ever. But in the spirit of the Russell-Einstein Manifesto, we now call on all scientists to expand our concerns to a broader set of interrelated dangers: destruction of the environment on a global scale and denial of basic needs for a growing majority of humankind. Without reducing our commitment to arms reduction and war prevention, we must recognize that environmental degradation and large-scale impoverishment are already facts and can lead to massive catastrophe even if nuclear war is avoided.

The present inequitable international economic order confines many countries to the crushing cycle of poverty and induces them to use environmentally destructive industrial and agricultural practices. When coupled with worldwide population growth, and excessive production and profligate consumerism in the industrial nations, this is pushing the planet toward disaster.

Today's pattern of increasing energy use is a key link in a dangerous web of international environmental problems. Among these are global climate change, ozone depletion, acid deposition, and water pollution. These, combined with other potentially catastrophic effects, including deforestation, soil erosion, and mass extinction of species, reduce the earth's ability to support a growing population. The combined effect diminishes ecosystem functions in ways that will damage economies in the North and fatally undermine economies in the South.

These linked environmental problems affect all nations. They exacerbate international tensions and increase the risk of future conflicts through the impacts of sea-level rises, forced migrations, and persistent crop failures.

To survive, we must recognize that ënvironmental degradation weakens the security öf all. The challenge is to fund ways to promote sustainable development of all regions in the world while reducing both military and ecological threats. Cooperation among nations, and effective organisations at the international, national, regional and local levels, are essential to maintain earth's life-support systems. Intense efforts must be made to foster a feeling of connectedness and cooperation and to correct economic injustices and promote trust.

The steps taken up to the present to halt environmental destruction have proved inadequate. Much stronger measures are required now. These include the development of alternative high-yield agricultural methods, while recognizing the value of some traditional practices, in order to conserve scarce water and topsoil. They will also entail strict regulation of industry and land-use, and massive investment in environmentally sound practices, increased efficiency of resource use, deployment of renewable energy technologies, poverty reduction, and population planning. Education must promote a shift toward lifestyles compatible with the preservation of our life-support systems. Global use of fossil fuels must be reduced. The 1987 Montreal Protocol on Ozone Depletion must be strengthened to eliminate the production and use of chlorofluorocarbons. International support for reforestation must be increased dramatically. In this way, the planet may move toward a new and stable balance in which nature can withstand the impacts of human civilization.

Acronyms

ABM	anti-ballistic missile
AEC	Atomic Energy Commission
ALCM	air-launched cruise missile
ASAT	anti-satellite weapon
ASW	anti-submarine warfare
BMD	ballistic missile defence
C^3I	command, control, communications, and intelligence
CBM	confidence-building measure
CCA	Commission for Conventional Armaments
CDE	Conference on Disarmament in Europe
CPSU	Communist Party of the Soviet Union
CSCE	Conference on Security and Cooperation in Europe
CSBM	confidence- and security-building measure
CST	Conventional Stability Talks
CTB	comprehensive test ban
CW	chemical warfare
DOE	Department of Energy
ECM	electronic counter-measures
EMP	electromagnetic pulse
ENDC	Eighteen Nation Disarmament Committee
ESA	European Space Agency
FBS	forward-based system
GCD	general and complete disarmament
IAEA	International Atomic Energy Agency
IDC	International Data Centre
IRIS	International Research Institute for Seismology
IRM	intermediate-range missile
ISMA	International Satellite Monitoring Agency
kkv	kinetic kill vehicle
LWR	light water reactor
LYTB	limited yield test ban
MBFR	mutual and balanced force reduction
MBT	main battle tank
MLRS	multiple launcher rocket system
NATO	North Atlantic Treaty Organization
NDC	National Data Centre
NPT	Non-Proliferation Treaty
NRDC	Natural Resources Defense Council
NTM	national technical means
OTA	Office of Technology Assessment
PGM	precision-guided munitions
PPM	perimeter and portal monitoring
RDT&E	research, development, test and evaluation
SALT	Strategic Arms Limitation Talks
SBI	space-based interceptor
SCC	Standing Consultative Commission
SDI	Strategic Defense Initiative

SIGINT	signals intelligence
START	Strategic Arms Reduction Talks
UNCSTD	United Nations Conference on Science and Technology for Development
UNCTAD	United Nations Conference on Trade and Development
UNIDIR	United Nations Institute for Disarmament Research
WEU	Western European Union
WTO	Warsaw Treaty Organization

Notes on the Contributors

Kirill Babievsky (USSR) organic chemistry. Professor of Organic Chemistry at the Institute of Organoelement Compounds, Moscow.

Nicole Ball (USA) economics. Director of Analysis, National Security Archives, Washington DC

Valerii Davydov (USSR) arms control. Institute of the USA and Canada Studies of the USSR Academy of Sciences, Moscow.

Peter Deak (Hungary) military science. Colonel. Head of Science Department, Ministry of Defence, Budapest.

Anne Ehrlich (USA) population biology. Senior Research Associate, Stanford University, California.

Paul Ehrlich (USA) biology. Bing Professor of Population Studies, Stanford University, California.

Nikolai Enikolopov (USSR) chemical physics. Director of Institute of Synthetic Polymer Materials, USSR Academy of Sciences, Moscow.

Esmat Ezz (Egypt) chemical and nuclear defence. Retired Major-General. Director of Scientific Research Branch, Cairo.

Harold A Feiveson (USA) arms control. Research Scientist at Center for Energy and Environment Studies, Princeton N.J.

Richard L Garwin (USA) physics. IBM Fellow and Adjunct Professor of Physics at Columbia University, New York.

Peter H Gleick (USA) environment and security. Director, Global Environment Program, Pacific Institute, Berkeley, California.

Mikhail Gokhberg (USSR) physics. Professor of Physics, Institute of Earth and Physics, USSR Academy of Sciences, Moscow.

Vitalii Goldanskii (USSR) physics and chemistry. Director of Institute of Chemical Physics, USSR Academy of Sciences, Moscow.

Klaus Gottstein (FRG) elementary particle physics. Director of Research Unit, Max Planck Society, Munich.

Andrew Haines (UK) epidemiology. Professor of Primary Health Care, University College and Middlesex School of Medicine, London.

John P Holdren (USA) energy, environment and arms control. Professor of Energy and Resources, University of California, Berkeley.

Erhard Keppler (FRG) environment. Technical Director of Max Planck Institute for Aeronomy, Kaltenburg-Lindau.

Jean Klein (France) arms control and disarmament. Institute Francais des Relations Internationales, Paris.

Jeremy Leggett (UK) geology. Director of Science, Greenpeace, London,

Mikhail Milstein (USSR) military policy and arms control. Retired General. Section Chief at the Institute of USA and Canada Studies of the USSR Academy of Sciences, Moscow.

Albrecht A C von Müller (FRG) European security policy. Director of Research Programme, Max Planck Society, Starnberg.

Julian Perry Robinson (UK) science policy. Senior Fellow, Science Policy Research Unit at University of Sussex, Brighton.

John Pike (USA) space policy. Director for Space Policy, Federation of American Scientists, Washington DC.

Judith Reppy (USA) economics and peace studies. Associate Director of Peace Studies Program, Cornell University, Ithaca, NY.

Roger Revelle (USA) science and public policy. Professor of Science and Public Affairs at University of California, La Jolla.

Joseph Rotblat (UK) physics. Emeritus Professor of Physics, University of London. President of Pugwash, London,

Adam Rotfeld (Poland) European security and East-West relations. Head of the European Security Department at the Polish Institute of International Affairs, Warsaw.

Jack Ruina (USA) electrical engineering. Professor of electrical engineering at Massachusetts Institute of Technology, Cambridge, MA.

Maurice Strong (Canada) environment. Strovest Holdings Inc., Ottawa.

Krishnaswami Subrahmanyam (India) national and international security issues. Nehru Fellow, at the Institute for Defence Studies and Analyses, New Delhi.

Ronald G Sutherland (Canada) arms control and disarmament. Department of External Affairs, Ottawa.

Theodore B Taylor (USA) physics. Former nuclear weapons designer at Los Alamos National Laboratory. Consultant, West Clarksville NY.

Bhalchandra M Udgaonkar (India) theoretical physics. Senior Professor at Tata Institute of Fundamental Research, Bombay.

Caesar Voute (Netherlands) geology. Former Professor in General and Applied Geology at International Institute for Aerospace Survey and Earth Sciences, Enschede.

Subject Index